Evolution as Entropy

•

Science and Its Conceptual Foundations
David L. Hull, Editor

Evolution as Entropy

•

Toward a Unified Theory
of Biology

Daniel R. Brooks
and
E.O. Wiley

The University of Chicago Press
Chicago and London

Daniel R. Brooks is associate professor of zoology at the University of British Columbia. He is the 1985 winner of the Henry Baldwin Ward Medal of the American Society of Parasitologists. E. O. Wiley is associate curator of the Museum of Natural History and associate professor of systematics and ecology at the University of Kansas. He is the author of *Phylogenetics, the Theory and Practice of Phylogenetic Systematics* and cocditor of *Zoogeography of North American Freshwater Fishes*.

The University of Chicago Press, Chicago 60637
The University of Chicago Press, Ltd., London

© 1986 by The University of Chicago
All rights reserved. Published 1986
Printed in the United States of America

95 94 93 92 91 90 89 88 87 86   5 4 3 2 1

Library of Congress Cataloging-in-Publication Data

Brooks, D. R. (Daniel R.), 1951–
  Evolution as entropy.

  Bibliography: p.
  Includes index.
  1. Evolution.   2. Entropy.   I. Wiley, E. O.
II. Title.
QH371.B69   1986   575.01'6   85–8544
ISBN 0–226–07581–8

· Contents ·

· *Preface* ·

That organisms have evolved rather than having been created is the single most important and unifying principle of modern biology. Theories regarding the causal mechanisms of evolution are not so important in "proving" its reality. The fact that scientists put forward theories means that they accept this reality. Confused creationists frequently think that if they can "disprove" Darwin's theory of natural selection they can "disprove" evolution. But of course this is untrue—even if they succeeded they would only be disproving *a theory* and not *the process*. The importance of such theories lies in the fact that theories provide a framework for interpreting the results of the evolutionary process. Thus, any theory of importance should be closely scrutinized because it affects the way evolutionary biologists conduct their research.

In this book we will develop the idea that evolution is an axiomatic consequence of organismic information and cohesion systems obeying the second law of thermodynamics in a manner analogous to, but not identical with, the consequences of the second law's usual application in physical and chemical systems. By "axiomatic" we mean that the results are necessary consequences or outcomes. By organismic information we mean genetic systems, cytoplasmic organization, and chromosome organization. By cohesion systems we mean systems of reproduction, systems that either integrate organismic information above the individual level of organization (sexual systems of reproduction) or fail to do so (asexual systems, or the splitting of sexual gene pools).

If evolution is an axiomatic consequence of certain biological processes following the second law, then current theories of the evolutionary process must necessarily be incomplete because they are theories of proximal cause. Biologists have traditionally associated evolutionary notions of axiomatic behavior or "ultimate cause" with various sorts of teleological behavior (e.g., orthogenesis) or theistic design. In a real sense Darwin produced a theory of proximal cause, evolution by natural selection, designed to exorcise the specter of supernatural design. In doing so, he moved biology into a Newtonian framework. Never mind how the ball began rolling; let us investigate the reasons it changes its speed or trajectory. Theories such as evolution by natural selection have the status of auxiliary laws; they are constraining laws that serve to select from the set of all possible events governed by a natural law the behavior of the particular phenomena observed. Constraints define what is actual or possible. We will develop the idea that the major constraints placed on evolution are not environmental but historical, that the direction of evolution is, in large part, determined by the past history of species and that natural selection and other proximal factors are primarily rate-limiting and not direction-giving constraints.

Scientific conceptual frameworks are often constructed as a core hypothesis and associated auxiliary hypotheses. The core hypothesis tends to be a simple and concise statement tendered as the fundamental truth encompassed by the conceptual framework. Two examples are "$E = mc^2$" from relativity theory and "earth and life evolve together" from panbiogeographical and vicariance biogeographical theory. Our core hypothesis is "biological evolution is an entropic process." The consideration that led us to this core hypothesis is presented in chapter 2. Core hypotheses are important because they suggest a unified perspective for a variety of research programs. On the other hand, the simplicity of core hypotheses makes them difficult to test directly, and the research programs unified by the core hypothesis form a set of auxiliary hypotheses that tend to "buffer" the core hypothesis from direct testing. Since the auxiliary hypotheses are drawn from research programs, they should be highly amenable to direct testing. Thus, most of the testing of any new general theory involves the auxiliary hypotheses. In chapters 3 to 6, we present auxiliary hypotheses in developmental biology, population biology and speciation theory, systematics, and community ecology.

Two concepts figure prominently in our formulation of the core hypothesis: *entropy* and *information*. These concepts are important for three reasons: (1) they provide a connection between biological processes and natural physical laws showing that biological systems are not governed by special laws of biology, (2) they provide a means for demonstrating the plausibility of nonrandom internally driven evolutionary change, and (3) they provide the conceptual link for the auxiliary hypotheses. In this book, we will not be extending entropy or information beyond these three uses. Those readers who are familiar with the history of information theory will be aware of the technical difficulties involved in many applications of information theory to biology. Our applications avoid those pitfalls, mostly by being conservative about what we expect measures of entropy and information to provide.

The theory we have proposed (Wiley and Brooks 1982, 1983; Brooks and Wiley 1984), and will further develop, is the outcome of our being systematists who adhere to a particular methodological approach called *phylogenetic systematics*. The founder of this approach, the late German entomologist Willi Hennig, was interested in formulating a "general reference system" for classifying organic diversity. His choice was a system based on genealogy. Hennig reasoned that no matter what proximal changes organisms or species might experience, the one thing that would never change is their genealogies. In addition, Hennig recognized that genealogical structure is inherently hierarchical. This was our first clue that history could behave in an axiomatic manner. In the past ten years, we and other phylogeneticists have devoted much of our primary research efforts to discovering historical effects in developmental patterns, comparative morphology, functional morphology, population and speciation patterns, biogeography, and ecology. Such historical effects seem ubiquitous. We began to believe that these historical effects were *constraints* on the evolutionary process, suggesting the influence of a natural law of history.

Our attention was progressively drawn to discussions of inherent order in development (e.g., Løvtrup 1974) and comparative morphology (e.g., Riedl 1978). An article by Farris (1979) concerning the information content of phylogenetic systematic analyses led us to information theory. When it was discovered that phylogenetic systematic techniques select the minimum entropy configuration of information in a set of observations about organisms (Brooks 1981a), our search focused on finding a connection among history, information, and minimum entropy configurations.

We believe that a deeper understanding of evolution lies in understanding the nature of entropy production, or entropy increases, in nonequilibrium systems. We consider the concept of time-dependent behavior embodied in the second law of thermodynamics to be more general than the concept of energy flows for which the law was originally formulated. In chapter 2, we attempt to show that the expected outcome of historical constraints on the action of the second law in biological systems is self-organization. The axiomatic behavior of living systems should be increasing complexity and self-organization *as a result of, not at the expense of,* increasing entropy.

In subsequent chapters, we explore this deduction for various levels of biological organization, from ontogeny (chapter 3), to populations (chapter 4), to phylogeny (chapter 5), and to the organization of communities (chapter 6). We wish to show that increases in complexity, organization, and entropy can be quantified at these functional levels and that our theory allows the operation of proximal causes embodied in Darwinism and neo-Darwinism without significant modification of the theory of natural selection. Our intent is to show that many well-understood evolutionary processes can be easily subsumed by our more general theory of evolution and that what is popularly known as neo-Darwinism will retain its place as a relatively complete theory of proximal cause in biology. We hope to show more reductionist-minded workers that there are phenomena of relevance and interest in higher functional levels (populations, species) among biological systems. By the same token, we hope to show biologists that adoption of a more general theory of evolution, one with direct links with physics, does not entail a reduction of biology to atomistic physical principles. There is good physics in biology, and good biology in physics; cross-fertilization should make both fields more productive. We would like nothing better than for this book to aid in taking a step toward real unification.

· *Acknowledgments* ·

We have greatly benefited from discussions with many friends and colleagues whose frank criticisms and willingness to talk about new ideas we greatly appreciated. They include Cyril V. Finnegan, G. G. E. Scudder, Jack Maze, Lionel Harrison, David Holm, Paul LeBlond, D. David Cumming, Richard T. O'Grady, and Douglas Begle, all at the University of British Columbia; Robert D. Holt, Robert S. Hoffmann, Philip S. Humphrey, James L. Hamrick, Richard L. Mayden, Darrel Frost, Havid Hillis, and Kathleen Shaw at the University of Kansas; Joel Cracraft (University of Illinois); John Lynch (University of Nebraska, Lincoln), James Dale Smith, William F. Presch, David Depew and Bruce Weber (California State University, Fullerton); Richard Pimentel and Ronda Riggins (California Polytechnic and State University, San Louis Obispo); Vicki A. Funk (U.S. National Museum of Natural History); Virginia R. Ferris (Purdue University), Steven Frautschi (California Institute of Technology); Stuart Kauffman (University of Pennsylvania); John A. Endler (University of Utah); Jane Maienschein and James Collins (Arizona State University); James H. Brown (University of Arizona); John Collier (University of Calgary); Lionel Johnson (Freshwater Institute, Winnipeg); and Roger Hansell (University of Toronto). We would also like to thank the many faculty members and graduate students we have met and talked with during our visits to the Los Angeles County Museum, the University of Oklahoma, the California State University at Fullerton, the University of Nebraska–Lincoln, the University of Toronto, the University of New Orleans and the Louisiana Nature Center, the University of New Mexico, the University of Utah, California Polytechnic and State University, San Louis Obispo, Harvard University, and the University of Texas, Austin.

In addition to helpful discussions, we are most grateful to the following persons. George V. Lauder (University of Chicago) and John Collier read the entire manuscript. Sally Frost (University of

Kansas) read chapter 3. Tony Joern (University of Nebraska–Lincoln) read chapter 6. Their comments are greatly appreciated, although not always followed.

We thank the three anonymous reviewers who read the manuscript for the University of Chicago Press for their detailed and highly instructive criticisms and recommendations for revision.

Four individuals published papers that significantly influenced the development of our ideas and each has been willing to discuss his ideas, and our differences, on several occasions and through correspondence. They are Jeffrey Wicken (Behrend College, Pennsylvania State University), George Karreman (University of Pennsylvania), David Layzer (Harvard University), and Peter Landsberg (University of Southhampton and the University of Florida).

David L. Hull (Northwestern University) has supported our efforts from the beginning. We have valued his comments, his criticisms, and his attitude that new ideas should have a forum.

Maggie Hampong illustrated the book for us. Kathy Gorkoff did the word processing. Their skills are greatly appreciated.

Many of our ideas and some of the manuscript were developed while we were being supported by the Natural Sciences and Engineering Research Council of Canada (NSERC Grant A7696 to DRB) and the National Science Foundation of the United States (NSF Grant DEB 8103532 to EOW). While neither agency supported this project per se, their willingness to support our other research has been instrumental to our scientific development, and we thank both agencies for their support. We also thank the University of British Columbia and the University of Kansas for their support and for providing intellectual environments that have materially contributed to our efforts.

Finally, we thank Grace Karreman and Karen Wiley for their moral support, good humor, and numerous sacrifices throughout the course of this project.

· 1 ·

Prelude

In examining what we thought was the current evolutionary synthesis, we were reminded of Alice's experience with the Cheshire cat; the harder we looked, the harder it was to see any "synthesis." This is not meant as a criticism per se, but it does imply a very diverse pluralism on the part of evolutionary biologists and a lack of any unifying theory that ties together the diverse threads of evolutionary research. It would be easy to pretend that neo-Darwinism, for example, represented such a theory. But such a pretense would result in a caricature of a theory that does not, in our minds, exist. Rather than erect straw men, we feel more comfortable outlining, albeit incompletely, what we perceive as the current state of evolutionary theory, what attitudes toward the process of evolution are held, and what we think is missing. The positive aspects of our current knowledge of the evolutionary process are summarized below.

1. It is generally acknowledged that Weissman was correct; heritable characters are transmitted through the germ plasm and Lamarckian mechanisms cannot account for the major features of the evolutionary process.

2. Heritable characters are transmitted to offspring via genes, cytoplasmic organization (a manifestation of the genes of the previous generation), and chromosomal organization. Phenotypes are manifested through various interactions of genes, cytoplasm, and environmental effects during development and growth.

3. Heritable characters may be changed by a variety of mutational events. The earlier in development a mutation occurs, the greater its effect but the less its probability of being successful (i.e., of resulting in a viable and fertile individual).

4. A body of knowledge, population genetics, can be applied to explain how particular genotypes and/or phenotypes behave in populations. The principles of population genetics provide a link between the changes that occur in individuals and the fates of these changes in populations and species.

5. A body of knowledge, population ecology, can be applied to explain the dynamics of populations in particular environmental circumstances. The principles of population ecology provide a link between population genetics and the environment.

6. Species may be formed from previously existing species, and the result is either an increase in the number of species (the usual outcome) or a decrease in the number of species (a theoretical possibility).

7. The physical separation or near separation of the gene pool (collective populations of a species) of an ancestral species seems to be the basis for most speciation, but other modes of speciation are possible.

8. Species are related to each other in a hierarchical manner and this is also manifested in their characteristics.

9. Certain groups of species show correlated patterns in time and space that result in common biogeographical patterns.

This admittedly incomplete list of evolutionary knowledge points to the fact that much is known about the processes at work in evolution. However, there are still many things we do not know, and many of the things we do know lack integrative principles to tie them together. Many of the controversies in evolutionary biology (neutralism vs. selectionism, the relative importance of competition, gradualism vs. punctuated equilibrium) concern the relative importance of phenomena rather than whether the phenomena are real. Some problems are generated because of "labels." For example, "neo-Darwinism" and the "synthetic theory" are popular labels, applied by those who generally accept that natural selection plays the most important role in the evolutionary process. "Opponents" of neo-Darwinism have a hard time attacking the "theory" because every time they do they are informed that their ideas are quite compatible with the "synthesis" (e.g., see the answers to various critics by Stebbins and Ayala 1981; Charlesworth, Lande, and Slatkin 1982). Yet it is not so important to say that some phenomenon, such as developmental constraints, is compatible with one's label. Rather, one must demonstrate that one's theoretical framework has incorporated mechanisms that can be used to explain the phenomenon.

We are dissatisfied with the current state of evolutionary theorizing, not because we think it is all wrong, but because (1) we do not believe that a truly integrated theoretical framework has been developed and (2) there are certain aspects of the evolutionary process that have yet to be integrated into any framework. We recognize four major items of unfinished business.

1. Evolutionary theory has never fully come to grips with the underlying causal laws of chemistry and physics.

2. Developmental biology has not been successfully integrated into the theoretical framework.

3. Existing evolutionary theory has failed to provide a rationale for the existence of higher taxa (groups of species produced by descent) that is consistent with our knowledge of phylogeny and population genetics.

4. Existing evolutionary theory has failed to provide what we would consider to be a robust explanation of the relationship between form and function in evolution.

We do not believe these shortcomings are resolvable within the current theoretical framework. We seek to provide a new framework, one that will incorporate certain parts of the old framework and include those aspects of evolution not now adequately explained. We will now examine these shortcomings in some detail.

Physical Laws and Evolutionary Theory

Modern physics and physical chemistry make a distinction between *microscopic* and *macroscopic* processes. Microscopic processes are time-independent, or reversible, whereas macroscopic processes are time-dependent, or irreversible. If one filmed a system undergoing microscopic changes and then showed the film to a second person, that second observer could not tell, from the film alone, if it were being shown in correct or in reverse temporal order. By contrast, macroscopic processes provide a system with an "arrow of time" that strongly distinguishes the "correct" or "natural" sequence of events from its reverse.

Newtonian processes are microscopic, and Darwin apparently intended his theory of natural selection to be consistent with Newtonian causality (Ruse 1982). Contemporary population biology, the primary source of evidence concerning the effects of natural selection, continues to treat biological evolution as a microscopic process. For example, if one were shown a film of a population of peppermoths changing from predominantly black to predominantly white, correlated with changes in the amount of industrial pollution (an independent environmental variable), one could tell that natural selection had occurred, but one could not tell whether the events were being shown in the correct or in reverse order of their actual occurrence. This tells us that the theory of natural selection is a microscopic theory, one having no arrow of time. Maynard Smith (1970) claimed that the modern formulation of the theory of natural selection, including Fisher's "fundamental theorem of natural selection," could not be construed as providing an arrow of time for biological evolution despite Fisher's claim to the contrary.

There are aspects of biological functioning that seem to be highly time-dependent, such as reproduction and development. One only need see a picture of an adult organism and a juvenile of any species to be able to tell the correct order of transformation. Most significant to us, though, is the recognition that the evolution of biological diversity seems to be a macroscopic phenomenon. The evolution of mammals, for instance, would appear very different if it were viewed in reverse historical order. Biological evolutionary theory will never be even reasonably complete until it includes a concept of macroscopic causality. This would simultaneously connect biological evolution with physicochemical evolution and provide an arrow of time for biology. As we will discuss more fully in chapter 2, there is only a single macroscopic law—the second law of thermodynamics. It is the law with which biology must be reconciled.

The second law of thermodynamics may be stated in a variety of ways, some more useful than others. In a general sense, it asserts that the universe as a whole or any isolated part of the universe is moving toward maximum entropy given the constraints operating on the system (Morowitz 1968). In terms of energy flow, this process is irreversible: the amount of free energy of the whole system decreases with time. Thus the second law is associated with the concept of time and history (Blum 1968). On a more concrete level, the second law has a statistical aspect. We can say that a system will become more randomized over time or that the number of states a system may occupy will increase until there is an equal probability that any one part of the system will be in any of the states available to it. When this occurs, the system is in equilibrium. Systems that are not in equilibrium may be moving toward equilibrium, or they may maintain themselves some distance from equilibrium by processing free energy. The "distance" from equilibrium is a manifestation of the order and organization of a system.

Those who have approached the problem of the relationship between the second law and evolution have come to a variety of conclusions. Lotka (1924) essentially synonymized the second law and evolution because both were concerned with the history of irreversible changes. This attitude was shared by other workers of the early twentieth century (see Lotka 1924, 1956 reprint, 26, 357). Others such as Bridgeman (1941) and Schrödinger (1945) sensed a potential conflict between the second law and evolution, because while the direction of change in the universe should be toward disorder and increasing entropy, the direction of change in evolution seemed to be toward greater order and decreasing entropy. Schrödinger (1945) explained this apparent contradiction by suggesting that the very existence of living systems depends on increasing the entropy of their local surroundings. Thus there can be a balance—the decreased entropy representative of life compensated for by increased entropy of the universe as a whole. The second law is not violated but only locally circumvented at the expense of a global increase in entropy (see Morowitz 1968 for a review of these views).

Prior to World War II, thermodynamic theory was concerned with equilibrium systems. After World War II, the discipline of nonequilibrium thermodynamics was developed. Prigogine and Wiame (1946) searched for the relationship between nonequilibrium thermodynamics and biological development. Obviously, organisms are nonequilibrium systems. Prigogine's group as well as Morowitz have attempted to find the analogies between organisms and nonliving nonequilibrium systems.

What has characterized this work is a preoccupation with the thermodynamic aspects of the evolutionary process. As we will show in chapter 2, all of these attempts are unsatisfactory for the paradoxical reason that a thermodynamic basis (= energy flow or flux) was suggested for evolution rather than an attempt to extend the second law to the very attributes of organisms that make them different from nonliving physical systems: their genes, cytoplasmic organization, mating systems, and how they speciate. The major thesis we will present is that evolution is an entropic phenomenon, but because the levels of organization to which the second law applies in a causative manner in biology are levels that are absent in atomistic physical systems, biology cannot be reduced to atomistic physics.

Evolution and Developmental Biology

Biologists have been interested in two aspects of the development of organisms for at least 190 years. First, what is the reason for the holistic organization, integrated action, and characteristic sequence of events seen in the ontogenies of organisms? Second, what is the reason that the ontogenies (or at least some stages in the ontogenies) of different species resemble each other closely? Two schools of thought, one a "recapitulation" school and one a "differentiation" school have dominated biology, first from a nonevolutionary perspective and then from an evolutionary perspective. In almost all cases, both of the questions posed have been considered together.

The doctrine of recapitulation was first advocated by preevolutionists such as Kielmeyer, Meckel, Oken, and Serres. Holmes (1944) discussed the views of this group of biologists and concluded that they developed no plausible explanation of why recapitulation might occur. However, they all held that each stage in development represented a previous adult form and that similarities in ontogenies of different species represented similar "affinities" of the adults. Those affinities ranged from proximity to each other in the natural scale of life created by a deity to similarities in the natural laws responsible for their particular "creation" by spontaneous generation.

An alternative view was espoused by von Baer (1828), who suggested that the parallelism between adult forms and developmental sequences indicated progressive differentiation from a similar beginning point rather than recapitulation of similar past adult stages. Von Baer's views were based on Aristotelian essentialism, or conformity to type: the "telos," or goal, of an embryo was the adult form. Similar teloi acting on similar starting products produced similar ontogenies. His generalizations were derived from empirical observations that he characterized into four dicta: (1) The commonly shared features of a group appear before the special features. (2) Less general features develop from more general ones. (3) Rather than passing through adult stages of lower organisms, every embryo of a species diverges (differentiates) progressively from them. (4) Embryos of advanced species never resemble adults of a lower organism. Von Baer's views were echoed by Robert Chambers (1844), who bridged the gap between the evolutionary and nonevolutionary traditions by asserting that evolution is merely orderly creation and that phylogeny and ontogeny are more or less alike because the same creative forces operate in both. In both the nonevolutionary recapitulationist view and von Baer's view, species were immutable "types," units formed once and maintained by reproduction. The emerging recognition during the eighteenth and nineteenth century that the similarities among species formed a divergent hierarchy rather than a continuum or scale of perfection led to a variety of evolutionary viewpoints.

Darwin (1859) and later Haeckel (1866) revived the doctrine of recapitulation in an evolutionary context. With the advent of evolutionary thinking, the type concept was abandoned because types are immutable and evolution is a process of change through time. Thus, because of its historical past, the organism might have inherited information from recent or even distant ancestors. If natural selection is constantly drawing forth new forms, the history of life would be found recapitulated in the ontogenies of contemporary species. Darwin invoked the functional, that is, adaptive/selective, needs of the organism to explain why recapitulation should be the predominant form of ontogenetic sequence as well as to explain departures from recapitulation.

Darwin's protagonist, Ernst Haeckel, extended the doctrine of recapitulation to such an extent that he termed *evolutionary recapitulation* the biogenetic law. According to Haeckel's views (1866, 1877), ontogeny is a short recapitulation of phylogeny and, more important, phylogeny is the mechanical cause of ontogeny. Ancestral adult stages were thought to be repeated during the development of the descendants but were crowded into earlier ontogenetic stages, making ontogeny a shortened version of phylogeny. Because the recapitulated stages represent the species' history, Haeckel called them *palingenetic*. When early stages of a developing animal exhibited structures that no adult ancestor could have had (such as the placental membranes of some mammals), Haeckel concluded that the recapitulation of palingenetic stages had been held in abeyance and an adaptive, or *caenogenetic,* stage intercalated. Von Baer's ideas would pertain only to the caenogenetic modifications which, by virtue of their having been called forth by natural selection to fill an adaptive need, could not shed light on phylogenetic relationships. True, palingenetic, evolutionary novelties must appear first in the adult stage, and phylogeny would proceed by means of accreting new adult stages onto old ones. Thus, palingenetic changes could be subject to natural selection that shaped phylogeny.

During the latter part of the nineteenth century, two traditions in developmental biology seem to have developed. One tradition would continue to search for laws of individual ontogenies as the overriding mechanism producing natural diversity, whereas the other would view ontogenies as the source of variation on which selection would operate—thus, the orderliness of individual ontogenies reflected adaptations to selective pressures. The first, exemplified by His (1874), suggested that the proper goal of embryology (developmental biology) was the discovery of causal laws of embryogenesis and not the manufacturing of phylogenetic trees. His (1874), Hertwig (1906), and Sedgewick (1894, 1909, cited in Holmes 1944) showed that even very closely related organisms could be distinguished early in ontogeny and that evolutionary novelties did not always appear at the end of ontogeny. Thus, they suggested that recapitulation was falsified and searched for alternative mechanisms to account for ontogeny. In a sense, this represented a return to the empiricism of von Baer, within an evolutionary framework. Neo-Lamarckism and orthogenesis were late nineteenth-century products of this tradition (Bowler 1983).

The second research tradition that developed at that time was exemplified by the cell-lineage workers in the laboratory of C. D. Whitman at Woods Hole Marine Biological Laboratory. Our discussion of this group is based on Maienschein's (1978) study. However, we have applied our own emphases and interpretations to some of the material she used. The work of the Woods Hole group was also based on the notion that recapitulation was an inadequate explanation of ontogeny. In contrast to His, they viewed the question of the significance of development for evolution as crucial. However, their view of the twofold nature of ontogeny differed from that stated at the beginning of this chapter. They asked: (1) To what extent is an embryo subject to external pressures and adaptive at every stage of its development? and (2) To what extent is each embryo a historical being primarily reminiscent of its ancestors because it is the result of past evolutionary change?

It is apparent that this group adopted the Darwinian view of evolution but did not believe that it required recapitulation. The initial premise of the group may be found in an article by Whitman (1888) suggesting that any embryonic stage that did not resemble an ancestral adult was a secondarily adaptive form without phylogenetic significance. Thereby Whitman could allow not only Haeckel's caenogenetic events but also modifications, by natural selection, of larval stages already present. Whitman's students then proposed various ways in which this idea could explain their embryological findings.

Wilson (1891, 1892) argued that persistence of an ancestral form was due to adaptive needs and also suggested that modification of those forms was due to adaptive needs. Thus, like Darwin, he proposed that a single mechanism explained both recapitulation and departures from it. He proposed two classes of homology, illustrating his preoccupation with the functional needs of the adult:

1. *Complete homology*—Same embryological origin, same adult structure.

2. *Incomplete homology*—Different embryological origin, same adult structure.

Treadwell (1898) found Wilson's distinctions unsatisfactory and pointed out that most homologies are "incomplete" and that most "complete" homologies have a third form, that of "same embryological origin, different adult structure."

Wilson (1895) summarized his modified ideas thusly:

1. Comparative morphology rather than comparative embryology should provide the key to homologues.

2. The adult form must be the guide by which homologies and thus biological relations are assessed.

3. Careful comparative morphological studies, from the one-celled stage upward, would provide a trustworthy basis for a new criterion of homology bringing together anatomical and embryological data.

4. This will not be easy because some adult homologies may originate in different parts of germ layers or from different cells.

Conklin (1896) suggested that early stages of development were more important than later stages, and Lillie (1898) suggested that the entire ontogeny was the unit that should be considered. In summarizing the views of the group, Lillie presented the following set of ontogenetic transformations:

ABCD
ABCDE
ABCDEF
ABCDEFG

and suggested that they represented the Haeckelian view of evolution of ontogenies. By contrast, Lillie asserted that real ontogenies looked like this:

ABCD
A'B'C'D'E
A"B"C"D"E'F
A‴B‴C‴D‴E"F'G

and he took this as de facto evidence that each stage for each species had been modified independently by natural selection. This meant that in such cases larval stages could not be relied on to provide data about phylogenetic relationships. The fact that some general affinities could always be discerned despite such adaptive modifications was explained by recapitulation, since it was believed that the only new stages produced were new adult stages. And if ancestral adult stages were adaptive, they would be retained unmodified in the ontogeny of the descendant. Thus, rather than tossing away recapitulation theory, he simply allowed natural selection to override it.

Whereas the Woods Hole group believed that all developmental stages could change since they were affected by natural selection, Garstang (1922) proposed that there had been evolution at the level of zygotes, resulting in modified ontogenies and thus modified adult structures. Similar sentiments were expressed by Ekman, Franz, Fuchs, Nauch, Schindewolf, and Shurnway (see Holmes 1944). To these developmentalists, ontogeny is the mechanical explanation for evolution rather than the reverse. The search for causal laws of embryogenesis begun by von Baer and reinforced by His found its advocates in these workers, who considered ontogeny to be a causal sequence.

However, the notion that all larval and adult features, especially those that appeared to be adaptations to particular environments, were the result of zygotic changes conflicted with the emerging neo-Darwinian view. Naef (1917, 1931) was one of the chief exponents of a modified view of ontogeny as a causal sequence, which still gave focus to the emergent adaptive needs of later stages of development. He proposed (1917) the "law of terminal modification" as follows: "The stages are conservative in proportion as they appear earlier, and progressive in proportion as they appear later in the ontogenetic series." Naef (1931) expanded his views thusly: "Morphogenetic processes of an ontogenetic stage through whose modification a following stage arises are to be looked upon as phylogenetically older than those which grew out of them." This left room for additions at the end of the ontogenetic process, leading to recapitulation. Sewertzoff (1931) expressed similar sentiments about the relative ease of terminal vs. nonterminal additions, but also reemphasized that nonterminal changes could occur. He recognized three types of ontogenetic change, those occurring in early, middle, or late ontogeny, and suggested three kinds of phylogenetic findings based on each one. For terminal modifications, recapitulation would hold true; for middle modifications, recapitulation would hold true up to the point of alteration (a point echoed by Goldschmidt 1940); and for very early changes, no recapitulation would occur at all. Sewertzoff (1931) concluded: "The law of recapitulation and von

Baer's Law are not universally valid; they are entirely valid only for organs which have developed by means of additions to end stages."

Holmes (1944), in a review article rejecting recapitulation in the sense of phylogeny determining ontogeny, stated:

> All evolution in organisms is simply the manifestation of the one basic evolution of the Keimbaum, or continuous germ plasm. . . . Moreover, gene mutations have to fit into the ontogeny without too greatly disrupting its course. If they manifest themselves toward the end of it, when there are few or no further processes to be thrown out of gear, they would not be so likely to be eliminated by natural selection. . . . Our present knowledge of lethal mutations clearly indicates that the fatality of gene expression in very early development is exceedingly high. How much of this mortality is attributable to the nature of the mutations and how much to the period in which they act we do not know. A priori, it seems probable that the earlier the change, and therefore the greater the subsequent modifications entailed, the less likely it would prove to be advantageous. Early stages would seem therefore to be guarded against change. They do change, but the directions of adaptive modifications are relatively limited, if they do not prove injurious to the subsequent ontogeny.

Holmes' views embodied two assumptions. First, germ plasm is relatively undifferentiated and thus highly similar in all species. Therefore, if the embryos of two closely related species looked much different, we could still be certain that their germ plasm was highly similar. Second, he assumed that changes early in ontogeny were generally selected against, and when they were allowed by natural selection, they had little impact on the rest of ontogeny because early ontogenetic stages lack adaptive freedom.

The role of natural selection, or environmental factors extrinsic to the organism, in shaping ontogeny was given perhaps its greatest backing by de Beer (1958), who stated:

> When, therefore, we ask the question: Do the internal factors which are present in the fertilized egg suffice to account for the normal development of an animal? we are also questioning if heredity is solely responsible for the sequence of processes which constitute ontogeny. . . .
> It may be definitely stated that the internal factors which were inherited from the parents are *not* sufficient to account

for the development of an animal. To illustrate this all-important point, we may refer to the fact that ever since the Silurian geological period, roughly 300 million years ago, vertebrate animals had two eyes, as can be proved from a study of the fossils. Since (then, as now) there must be internal factors concerned with the production of two eyes, these have been transmitted to every generation for a very considerable period. But these factors are not self-sufficient, for if a few pinches of simple salt (magnesium chloride) are added to the water in which a fish (*Fundulus*) is developing, that fish will undergo a modified process of development and have not two eyes, but one. . . .

This viewpoint is reinforced a few pages later:

The action of the internal factors is to ensure that if the external factors are normal and do evoke any response in development and produce an animal at all, that animal will develop along the same lines as its parents. The internal factors are only a partial cause of ontogeny.

An additional refinement of this approach to developmental biology and evolution was represented by the work of C. H. Waddington. His attempts to find a selective explanation for epigenesis and the causal sequence of ontogeny may be summarized in the following (1966):

The first question to ask is not, what is the detailed nature of the components—we can hardly expect to get an answer to this—but rather, what stable states will survive a disturbing stimulus? . . . What we need is statistical mechanics comparable to the thermodynamics of physical theory. But we shall have to elaborate one that applies to open systems which do not conserve either matter or energy; whose final state is not determined by initial conditions; and in which entropy can decrease.

Polikoff (1981) summarized Waddington's views of ontogeny and evolution thus:

(1) Phenotypes are the manifestations of complex and coordinated systems under genotypic control, (2) The environment may be very directly involved in epigenesis, and (3) Development generally tends toward relatively definite

end states (e.g., a discrete organ), and there is a canalization of developmental pathways which buffers the developmental system from disruption by environmental and genetic abnormalities.

And despite Polikoff's assertions that "these epigenetic concepts immediately contradict the premises of the 'neo-Darwinian paradigm,' " points 2 and 3 above exemplify the pattern, begun by Darwin and carried forth by Lillie and de Beer, of using factors extrinsic to the organism ("environment," "natural selection") to explain both the orderliness of ontogeny and departures from it. Either phylogeny created ontogeny or natural selection created ontogeny. Waddington's major contribution was the notion that the search for such a causal mechanism must involve causal laws of physics and chemistry, especially those pertaining to open systems.

Most major recent studies of the relationship between ontogeny and evolution have concentrated on very early, and epigenetic, effects in embryogenesis. We will refer in more detail later to some of the work summarized in this part of our discussion. Løvtrup (1974) presented a treatise on epigenetic effects, summarizing findings that indicate embryogenesis may proceed on the basis of relatively few genotypic instructions. He asserted that because such changes occurred virtually independent of the effects of natural selection and had such a profound influence on the adult form, the evolutionary process could not possibly be a Darwinian or neo-Darwinian one. Rather, Løvtrup asserted that evolutionary change comes about through modifications of the epigenetic program. In essence, all else is window dressing. However, he did not provide a causal mechanism for such changes beyond the epigenetic equivalent of random mutations and trial-and-error experiments by the whole developmental program. Riedl (1978) added the notion of constrained, nonrandom "mutations" to these ideas, developing his thesis from considerations of systems analysis.

Recent attempts to provide causal theories of development (Goodwin 1982; Goodwin and Trainor 1983) and of the relationship between ontogeny and evolution (Saunders and Ho 1976, 1981; Ho and Saunders 1979) have concentrated on notions of self-organizing, complexity-generating, entropy-production minimizing nonequilibrium systems. Goodwin's views embody the notion of morphogenetic fields. In Goodwin's view, there may exist a periodic table of viable morphogenetic fields (see Goodwin 1982). Any one of them can be perturbed by sufficient extrinsic perturbations to cause a shift to one of the others. The periodic table is thus highly deterministic, deriving from some set of initial conditions and constrained by inflexible physicochemical laws.

Saunders and Ho based their views on two major observations: (1) there is no increasing fitness gradient along which evolution appears to proceed, but there does seem to be a gradient of increasing structural complexity, and (2) much increase in variation and evolutionary change seems to occur not when selective pressures are heavy, but when the influence of natural selection lessens. Coupling these observations with their knowledge of epigenetic phenomena, which requires a theory of integrated development rather than one of independent genes, Ho and Saunders suggested that the neo-Darwinian paradigm was not adequate to explain the evolutionary process.

Ho and Saunders' alternate formulation rests on the assertion that it is structural complexity and not fitness or organization that increases during evolution. However, the increase in complexity occurs in an organized manner. This means that it is necessary to find an "organizing force" that directs the evolutionary increase in complexity. The variations in phenotype upon which selection operates are nonrandom effects of organism-environment interactions during development. According to Ho and Saunders, it is the epigenetic system that "internalizes" information from the environment during development; that internalized information is then used to direct subsequent evolution. Saunders and Ho (1981) drew an analogy between their view of the evolutionary process and nonequilibrium thermodynamics, especially the principle of minimum entropy production, in postulating a principle of minimum increase of complexity in evolution.

Ho and Saunders appear to have extended Waddington's (1966) world view of the relationship between evolution and ontogeny, including a nonequilibrium thermodynamic component. However, in considering ontogenetic programs to be the result of environmental effects on developmental variation, they consider the causal agent of change to be fast acting (environmental flux) and the changing part to be slow (canalization of "internalized" information). And yet, deterministic aspects of dissipative structures may be those components which resist change as well as those which cause change. Thus, there is no clear distinction made in this view. No provision is made for the emergence and canalization of new variants in the absence of environmental change, and yet evolutionary change is viewed as accelerating in the absence or lessening of selective pressures. With regard to the latter aspect, we are not certain how one could speak of a lessening of selection when the environment does not "lessen" but only changes. Thus, it might be more appropriate to state that when there is a sudden shift in selective regimes, the rate of realized variability increases. That being the case, this view does not represent a radical departure from the neo-Darwinian paradigm. And finally, there is no causal mechanism for the production of variants other than an assertion that "the environment" may perturb development to such an extent that new forms emerge. Thus, as with previous thinkers, a single factor, the environment, accounts for both order and departures from it.

It seems to us that there have been, and continue to be, a number of different views of ontogeny and its significance. There have been nonevolutionary and evolutionary groups, recapitulationists and differentiationists, and there have been those who sought the explanation for ontogeny in phylogeny, those who sought the explanation for phylogeny in ontogeny, and those who sought the explanation for both in natural selection. The search for intrinsic laws of development largely disappeared from evolutionary biology once the focus of research became centered on population phenomena. This was not only a productive shift, as the past sixty years will attest, but it was necessary to counter the perception of supernatural purpose or design (either progressive or degenerative) inherent in many treatments of neo-Lamarckism and orthogenesis (Bowler 1983). If the laws of ontogeny were supposed to supersede population-level phenomena, the empirical support for population-level phenomena certainly falsified such laws. Alternatively, if they did not override selection, the "laws of ontogeny" would be simply a synonym for the "laws of inheritance"; thus, their effect would be simply to produce random variation upon which selection acts. Evolutionary biologists in general lost interest in inherent laws of ontogeny. However, as Bowler (1983) pointed out, "It is certainly conceivable that in the excitement generated by the synthesis of Mendelism and natural selection, some genuine problems may have been overlooked." We will suggest that the laws affecting ontogeny constrain and organize the options upon which selection operates. In this way, both individual and population-level phenomena contribute to the production of organized diversity. It is our opinion that selection has been called upon to do more than it can. More complete evolutionary explanations require that we identify the relative contributions of the individual-level and of the population-level phenomena to the production of diversity.

Neo-Darwinism and the Origin of Higher Taxa:
Of Moths and Mammals

One major difference between our views and those of such workers as Simpson (1944, 1953), Rensch (1959), and Mayr (1963) concerns the reality of what is variously termed *evolution above the
species level, transspecific evolution, quantum evolution,* or simply
macroevolution.[1] These workers view the problem of the origin of
families, orders, and genera as problems to be explained in terms
of population phenomena. Mayr (1963) wrote: "The proponents of
the Synthetic Theory maintain that all evolution is due to the accumulation of small genetic changes, guided by natural selection,
and that transspecific evolution . . . is nothing but an extrapolation
and magnification of the events that take place within populations
and species." Rensch (1980) put the problem in historical perspective
by acknowledging that such explanations were put forward in response to critics of the neo-Darwinian synthesis:

> [Remane] repeated Weindenreich's criticism that we do not
> yet know mutations that cause those characters which are
> typical of higher categories. . . . Several other authors . . .
> also doubted that the origin of totally different types of an-
> atomical constructions that characterize phyla, classes, and
> orders of animals and plants could be explained by the con-
> tinuous effects of mutation and selection.[2]

From these quotes, we may conclude that (1) higher taxa demand
an explanation apart from the origin of the ancestors of those taxa
and (2) such an explanation can be formulated in a manner paralleling
explanations employed by population biologists.

That higher taxa require an "explanation" transcending the origin
of the ancestor of the members of each taxon may seem curious to
some. Many biologists think that higher taxa require recognition,
not explanation, and that the origin of a higher taxon is fully explained by the origin of the ancestor and first member of the group.

1. Not to be confused with "*macroevolution,*" as discussed by such workers as
Eldredge and Cracraft (1980) and Vrba (1980), which concerns study of the dynamics
of speciation and extinction.
2. Rensch's use of the word *category* is probably synonymous with the word *taxon,*
as distinguished by Hull (1976). Interpretations of many neo-Darwinists' discussions
of higher taxa are clouded by the fact that a clear distinction is not made between
taxon, or group of organisms, and category, or ranks in the Linnaean hierarchy.

However, systematists such as Simpson (1953) and Mayr (1963) thought that taxa became higher in rank not as a result of a speciation event, but as a result of changes occurring after the establishment of an ancestral clade, or perhaps even as a result of the independent attainment of particular morphological adaptations in several different clades. Simpson (1953) stated:

> A higher category [taxon] is higher because it *became* distinctive, varied, or both to a higher degree and not directly because of characteristics it had when it was arising.

And in reference to early and middle Paleocene mammal genera (Simpson 1953):

> If we knew of no placentals after the middle Paleocene we would certainly place them in a single order. *As of then* their proper comparative categorical rank was in fact that of an order. They are placed in six different orders because we recognize in them ancestors and allies of what later became six orders.

What Simpson seems to be saying is that while a higher taxon might have originated as a species, its rank is determined by what happened after its origin. Mayr (1963) agreed:

> It is not true that a new category [= taxon] arises as an order, class or phylum. It arises as a new species and eventually becomes a new genus which we assign to a new order only because its subsequent descendants show the degree of distinctness and of discontinuity (after such extinction) which by convention is considered to signify ordinal ranks.

At this point the issues become confused. Are higher taxa recognized as taxa because they arise from single species but ranked as "higher" because of events that occur after their origin? If so, then what is there to explain? The characters that make them a "higher" taxon are not the same characters that make them a *taxon*. Did Rensch (1980) intend to explain the characters that make mammals a taxon or the characters that make the taxon Mammalia a taxon with the rank of "class"? If the former, then the explanation involves no more than the explanation of morphological change associated with a single speciation event. However, if this is what Rensch, Mayr, Simpson, and others (see Dobzhansky et al. 1977) mean, then they should speak not of "transspecific" evolution but of cladogenesis as the explanation for higher taxa. Then all classifications would be phylogenetic *sensu* Hennig (1966), that is, classifications consistent with the entities involved (see Wiley 1981b). We conclude these authors wish to explain those characters that make the taxon a class rather than, say, an order or supercohort.

A Proposed Analogy

If we are to explain the origin and rank of a higher taxon such as Mammalia as an "extrapolation and magnification" of processes working within populations and species, then we should be able to construct a valid analogy between the two levels. We will first construct a series of questions and explanations used by population geneticists and evolutionary ecologists to explain evolutionary change in *Biston betularia* (see Kettlewell 1955, 1961, 1965) and then use a parallel series of questions and answers to explain the origin of Mammalia. We do not pretend to suggest that our questions and answers concerning Mammalia correspond to any series of statements made by such workers as G. G. Simpson. We would suggest that our analogy is drawn correctly and that *if* the neo-Darwinian systematists had constructed their arguments in this way they would have concluded that "transspecific evolution" could not be an extrapolation and magnification of changes within species in the manner they envisioned. We begin with peppermoths.

Question 1: Does the existence of black peppermoths require an explanation?
Answer: Yes, because there are also white peppermoths.
Question 2: Why are there black peppermoths?
Answer: Because in every generation of white peppermoths a predictable frequency of white gene alleles mutate to produce dominant black alleles, rendering a percentage of the offspring black.
Question 3: Must we explain the relative frequencies of black peppermoths?
Answer: Only if this frequency varies from place to place. Since we observe such changes in frequency from place to place in England, we should seek an explanation.
Question 4: Why are black peppermoths common in some areas of England and very rare in others?
Answer: Because the frequency of black peppermoths is positively correlated with the presence of soot-covered trees. Natural se-

lection favors black peppermoths in such environments; thus an increase in frequency over that expected by mutation pressure alone is observed. In contrast, when trees are not soot covered, white peppermoths have the advantage, and black peppermoths may actually be lower in frequency than predicted on the basis of mutation rates.

We may now turn to Mammalia.

Question 1: Does the existence of Mammalia require an explanation?
Answer: Yes, because there are also animals that are not mammals.
Question 2: Why are there mammals?
Answer: Among nonmammals evolution produced mammals.
Question 3: Must we explain the relative frequency of mammals?
Answer: Only if that frequency changes in time and space. Since it
    does, then we must seek an explanation.
Question 4: Why does the frequency of mammals vary over time
    and space?
Answer: Because the frequency of mammals is positively correlated
    with a mammalian adaptive zone variously subdivided into smaller
    adaptive zones. Natural selection favors mammals in such a zone.

There are several problems with the analogy.
1. White peppermoths are not analogous to nonmammals. Their ontological status is different. White peppermoths are a finite class of peppermoths. Nonmammals are a nongroup, a universal class.
2. Black peppermoths are a class of peppermoths, but mammals are not a class of nonmammals. One could conceive of mammals as a class of animals, but then Mammalia would be a finite class and it could have multiple origins, just as black peppermoths have multiple origins from parents who are white peppermoths. In fact, Simpson (1953) treated mammals in this fashion:

> Paleontologists use the arbitrary criterion that a reptile became a mammal when a dentary-squamosal joint developed and the functional jaw movement ceased to be on the articular-quadrate joint. This line was probably crossed separately by at least five different lineages (leading to monotremes, multituberculates, triconodonts, symetrodonts, and pantotheres, although it is just possible that two or three of these early differentiated from a single crossing of the line: there may have been some other late Triassic-early Jurassic crossings with early extinction).

It is possible to treat Mammalia as a finite class in this manner. However, in doing so we must adopt some construct to serve as a definition of the class so that membership can be determined. In doing so, we must give up two things, explanations of origins of higher taxa based on single speciation events (i.e., their origins as evolutionary groups) and homologous characters as diagnoses of higher taxa. In place of these explanations we must substitute a class construct based on homoplasy (i.e., convergence or parallelism).

3. The relative frequency of black peppermoths may be correlated with a variable that is independent with respect to the peppermoths themselves. There would be soot-covered trees whether there were ever any moths to light on them. In contrast, the relative frequency of mammals cannot be correlated with an independent variable because the presence of a "mammalian adaptive zone" can only be detected because there are mammals now occupying the "zone."

It would appear that the analogy drawn between microevolution and "transspecific evolution" is an incorrect analogy. We suggest that it is incorrect for both ontological and epistemological reasons. Population geneticists work with finite classes (i.e., closed systems) of phenotypes and their dynamics. Higher taxa are not classes but historical groups of individuals (i.e., open systems) (Wiley 1980a, 1981a). Thus, the ontological status of the entities involved is not analogous, much less identical. The characters involved are also different, at least if one wishes to work with homologies. Historical groups are discovered by finding characters to demonstrate that the group arose as an individual species, synapomorphies. The black allele of *Biston betularia* arises independently as an entropic phenomenon, mutation. Not all black peppermoths are descendants of a common ancestral black peppermoth. To the extent that Mammalia is a natural taxon we can discover, its characters once diagnosed a particular species, and all mammals arose from this single species. Thus, the epistemology of pigment alleles and of synapomorphies as sufficient criteria for group membership is different. In the first case, a "polyphyletic" origin of the character is acceptable and explainable. In the second case, a "polyphyletic" origin of the character would render that character invalid as a character uniting the group in question.

We conclude that the part of the synthetic theory that purportedly provides the auxiliary principles and procedures for reconstructing phylogenies and constructing biological classifications is not adequate for these tasks. Indeed, this conclusion can be reached intuitively when one considers that neither the Darwinian revolution (i.e., the acceptance of evolution) nor the rise of neo-Darwinism affected the manner in which organisms are classified. We will consider these problems in greater detail in chapter 5 and show that the theory we develop embodies procedures for accomplishing both tasks.

Form and Function

To us, one of the most interesting problems in biology concerns the relationship between structure (form) and function. Living organisms must take up and dissipate energy in order to maintain their structural integrity, and their structure often narrowly determines what form(s) of energy can be utilized. Therefore, a simple reduction of form to function or of function to form would seem to be misplaced in attempting to explain form *or* function. We are interested in distinguishing two aspects of this problem. First, how do form and function contribute to aspects of biological diversity we study today? Second, are form and function *causally* important in explaining the *evolution* of that biological diversity? For purposes of this discussion, we will refer to theories suggesting that structure is primarily an emergent property of function as *functionalist* theories and theories suggesting that function is primarily an emergent property of structure as *structuralist* theories. Questions concerning the contemporaneous relationship of form and function are nonhistorical, whereas those addressing the issue of the causal origins of biological systems in terms of form and function are generally historical. There are thus four classes of explanatory focus provided by functionalist and structuralist perspectives cast in nonhistorical or historical terms.

Nonhistorical functionalism asks the question, "How does this work?" or perhaps "How well does this work?" The explanation comprises a description of the function or the functional efficiency of the organism, trait, population, species, or ecosystem. Much of contemporary ecology, functional morphology, and physiology is devoted to finding mechanistic descriptive explanations for biological functions. Nonhistorical structuralism addresses questions such as, "What does this look like?" or "How is this structure constructed?" Much recent research along these lines has come from systematics and from developmental biology (Goodwin 1982; see also chapter 2). Such studies attempt to find mechanistic descriptive explanations for regularities in biological form.

The attempt to find causal explanations for the *origins* of the biological systems described in functional or structural terms has, for more than a century, generally involved a historical or evolutionary component. The Darwinian tradition (Darwinism and neo-Darwinism) tends to be one of historical functionalism. Structural changes are frequently viewed as essentially random with respect to causal evolutionary explanations, which are derived from functional considerations. Of these randomly produced structural variants in a population, those which persist are the ones that are functionally most efficient in the environment at any given time; they are defined as the "fittest" or "best adapted." Proponents of this particular view may experience difficulties in explaining the apparent structural regularity exhibited by living and fossil organisms unless the notion of historical constraint is introduced. Thus concepts of canalizing and stabilizing selection are important components in the neo-Darwinian research program. O'Grady (1984) has suggested that by explaining the evolution of particular structures in terms of the functions they perform, neo-Darwinian explanations cannot actively exclude language that suggests particular structures evolved in order to fulfill or to take advantage of the function. O'Grady termed such language *adaptational teleology* and noted that it was a problem primarily because neo-Darwinists do not *actually believe* that evolution is teleological.

Two other biological theories, neo-Lamarckism and orthogenesis, have been overtly teleological. We consider neo-Lamarckism to have been consistently a theory of historical functionalism; the "self-conscious" goals of better adaptation provided the "mechanism" for this optimistic view. Orthogenesis might at first glance be seen as solely a theory of historical structuralism, a process in which the evolution of each lineage proceeds through a series of structural modifications leading toward the final form and extinction. We must not forget that the concept of adaptive zones originally came from orthogenesis. Thus, it seems that the proponents of this view of evolution as internally driven linear change were concerned with questions of form and function embedded within a historical context (for a fuller discussion of neo-Lamarckism and orthogenesis, see Bowler 1983).

Recently, a group of developmental biologists (see Goodwin 1982) have suggested that nonhistorical structuralism contains the only causal explanations for biological diversity. Proponents of this view assert that there are a limited number of realizable biological structures determined by physical properties inherent in living organisms. Evolution is simply the manifestation of these structures. One assertion following from this view is that biological diversity might conform to a periodic table of morphology as well as it conforms to a hierarchical classification. However, such an assertion, if it were made, would have little or no empirical support from comparative biology. Proponents of this view also have trouble explaining the degree of functional "fit" of organisms to their environments. Furthermore, by denying historical influences, advocates of this view can never hope to distinguish physically excluded form from historically excluded form, which would seem to hamper their search for physically determined regularities in form.

It is our view that all of these approaches are lacking in some aspect important to a unified view of biology. What we are attempting to develop in this book is a theory of historical structuralism that is not teleological and that does not deny or ignore the importance of biological functions.

Summary

Current evolutionary thinking does not reconcile biological evolutionary theory with the four areas of contention discussed in this chapter. Such a reconciliation has not been reached because evolutionists have tended to focus on natural selection as the primary organizing factor in biological evolution. Thus, a connection with (1) natural physical laws and (2) inherently orderly developmental dynamics has not been established, (3) the origin of higher taxa without regard for adaptive zones is excluded, and (4) the relationship between form and function in evolution is not sufficiently explained by neo-Darwinism. We will attempt to develop in subsequent chapters a theory of biological evolution that unifies the empirical core of neo-Darwinism represented by population genetics, population ecology, and speciation theory with these four major areas of "unfinished business." The format we have chosen is designed with this goal in mind.

· 2 ·
Why Entropy?

It is our belief that the path to a more unified theory of biological evolution lies in a useful characterization of the causal laws responsible for its occurrence. Biological processes at many functional levels are characterized by irreversible changes over time, or macroscopic behavior. The changes involve both the uptake and use of energy and the transformation of matter from one structural state to another. They are routinely characterized by episodes of increasing organization and complexity.

Layzer (1977) characterized the relationship between causal laws and phenomena as follows:

$$\text{laws} + \text{auxiliary conditions} \longrightarrow \text{phenomena}$$

The auxiliary conditions can be boundary, initial, or symmetry conditions. As physics is presently constituted, laws and symmetry conditions are invariant under simultaneous inversions of charge, parity, and time. Thus, they do not distinguish between past and future, that is, they are not inherently irreversible and are thus considered to be microscopic. Layzer (1977) noted: "The Second Law of Thermodynamics is an apparent exception to this statement. But the Second Law is purely macroscopic; it has no microscopic counterpart. Hence, the directionality of the Second Law must be a consequence of auxiliary conditions." This means that systems which exhibit macroscopic behavior must be affected by boundary or initial conditions that behave in a lawlike manner.

Entropy Systems

Prigogine and Stengers (1984) have recently reiterated the division of macroscopic systems into those whose behavior is caused by their boundary conditions, which they call *thermodynamic systems,* and those whose behavior is caused by their initial conditions, which they called *dynamic systems.* We can readily see that this dichotomy is an artificial one, at least for biological systems. The development of a multicellular organism, for example, is caused by both a genetic program present at fertilization and by the environment in which the organism develops. What is needed, for biological theory at least, is a more general view of the processes of organization through time. This requires that we find a common currency to link "thermodynamic" and "dynamic" aspects of biological systems at all functional levels.

We will refer to all physical systems that exhibit time-dependent (irreversible) changes as *entropy systems* (or *entropic systems*), because the irreversible changes in all such systems are characterized by increases in an abstract quantity called the entropy of the system. The predictable behavior of such systems may be determined by different capacities, but most of our knowledge of the general properties of such systems derives from the study of strictly thermodynamic systems in which energy flows from the boundaries of the system are the determinate capacity. In our parlance, they might also be called *energy entropy systems.* By asking the question, "What is the relationship between biological evolution and entropy?" we find it necessary to consider two general components. The first is that of the particular mode of entropic behavior exhibited by biological evolution and the second is the identification of the parts of the living world that are responsible for the entropic behavior.

In order to understand and appreciate the concept of entropy as we construe it, it is necessary first to have an understanding of the concepts of reversibility and irreversibility, equilibrium and nonequilibrium, and history. Hollinger and Zenzen (1982) discussed three criteria for thermodynamic reversibility/irreversibility. They are:

I. A process is reversible if it can be reversed by changing external influences, i.e., if the system is at or near equilibrium. (Most processes are reversible at the macroscopic level and irreversible at the molecular level.)

II. A process is reversible if it can be reversed (conceptually, in time) by reversing the motion variables. Or we may say that the process is reversible if it can be reversed by reversing time. (All processes are reversible at any level.)

III. A process is reversible if the process and its reverse both occur naturally. (Most processes are irreversible at the macroscopic level even though there is no such irreversibility in the life of a single, isolated molecule.)

Hollinger and Zenzen asserted:

In our opinion the major cause of confusion about irreversibility in thermodynamics arises from the failure to distinguish between two different ways of reversing a process. Consider, for example, a process in which heat is flowing from a hot body A to a cold body B,

A (hot) $\longrightarrow$ B (cold).

If we reverse the situation so that A is cold and B is hot, then the spontaneous direction of heat flow is reversed *spatially,*

A (cold) $\longleftarrow$ B (hot)

but that is not the same thing as allowing the heat flow to be reversed in the sense of (III) [above]

A (hot) $\longleftarrow$ B (cold).

If A and B are near thermal equilibrium, then the flow of heat is spatially reversible, reversible in the sense of (I), but it is still thoroughly irreversible in the sense of (III).

They then presented the following assessment:

Thermodynamic Reversibility: All processes described by thermodynamics are always irreversible according to criterion III, always reversible according to criterion II, and mixed according to criterion I. At equilibrium they are reversible (I) with no creation or destruction of entropy. Away from equilibrium in the allowed direction (III) they are irreversible (I) and create entropy (III).

Two examples will help illustrate these distinctions.

Hydrogen gas in a bottle (type 1). Consider some hydrogen gas injected into a bottle in compressed form. The gas molecules will expand and disperse to fill the bottle, releasing measurable amounts of energy as they do so. When the limits of dispersion (the walls of the jar) are reached, the gas no longer expands and no more net energy is released.

Hydrogen emitted from the sun (type 2). The sun emits a stream of highly energized hydrogen ions. They become progressively more dispersed spatially and lose their original energy. They achieve thermal equilibrium with their surroundings and are thus susceptible to reversible heat flows in the sense of criterion I, depending on the temperature of the surroundings in which they find themselves.

Both type 1 and type 2 systems may be termed *equilibrium systems* because they exhibit nonfluctuating, progressive change (irreversible change in the sense of criterion III) from a state of relative order (high intrinsic, or bound, energy) to a state of relative disorder (low intrinsic energy). The point at which the net increase in disorder (net heat loss) ceases is called *thermodynamic equilibrium*. For type 1 systems, equilibrium is reached when the boundaries are reached. For type 2 systems, equilibrium is reached when the components of the expanding system reach the same temperature as their surroundings. Type 1 systems are sometimes called *closed,* or isolated, systems and type 2 systems are called *open* systems based on the degree to which the surroundings participate in the dynamics of the system. We can make our type 1 system a type 2 system by breaking the jar.

The transition toward equilibrium for type 1 systems is often given as

$$\Delta F = \Delta E - T\Delta S$$

where F is the free energy, or that released by the change in the system, E is the bound energy, or that which can be given up by the system, T is the absolute temperature, and S is the entropy, an abstract term relating to the macroscopic order or disorder of the system. (Since both structural considerations and energetics can be expressed in terms of molecular motion, the units are comparable.) As the system approaches equilibrium at constant temperature and pressure, F decreases ($\Delta F < 0$) because there is progressively less energy being given up and S increases ($\Delta S > 0$) because there is less energy left in the system to hold the particles together ($\Delta E < 0$). At equilibrium, $F = 0$ and no more work can be done.

Type 2 systems never reach $F = 0$ as an equilibrium point, but rather equilibrate with their surroundings. The transition toward equilibrium for such systems is usually given as the state of minimum Gibbs free energy, according to the equation

$$\Delta G = \Delta H - T\Delta S$$

where G is the Gibbs free energy, H the enthalpy (heat content of the system; not to be confused with Boltzmann's H, a measure of *statistical entropy* used throughout this book), and T and S are as given in the first equation.

Spontaneous change in such systems occurs when $\Delta G < 0$ (i.e., when energy is being dissipated from the system). The most favorable conditions are $\Delta H < 0$, $\Delta S > 0$. But if $T\Delta S > \Delta H$, ΔH can be positive. One way to achieve this is to increase pressure while keeping the number of molecules in the system constant. Temperature (T) will increase although the entropy need not. Alternatively, if pressure and temperature remain relatively constant, heat content of the system can increase ($\Delta H > 0$) only if more molecules are added (i.e., the system *grows*), and this will occur spontaneously only if entropy increases ($\Delta S > 0$). Because equilibrium is defined by heat symmetry between the system and its surroundings, the equilibrium value of G may rise or fall depending on the temperature of the surroundings. Consequently, S may rise or fall, but this "reversibility" is according to criterion I of Hollinger and Zenzen. No net entropy is produced because the relationship between system and surroundings is a fluctuation around an equilibrium.

As a further illustration, consider two pans of water side by side, one containing hot water and one containing cold water. The cold pan warms and the hot pan cools. Heat transfer always goes from where it is hot to where it is cold. Hollinger and Zenzen (1982) discussed this sort of behavior under the notion of "exclusion irreversibility."

> The processes which do occur in nature seem always to have increasing entropy (for the system and its surroundings), and they seem to be able to go only one way. If we believe that all processes are reversible in the sense of criterion II, then we may add that the excluded processes are those attended by decreasing entropy. There are many examples of this kind of irreversibility: heat flows from where it is hot to where it is cold, and never the other way, even though the other way is easy to conceptualize in terms of molecular motions; a chemical reaction on one side of equilibrium seems to have to go to that equilibrium, never the other way; the entropy of the world always increases.

While the pans are equilibrating, heat flow is in one direction only and entropy is being produced. At equilibrium, heat flow may proceed in either direction randomly and no entropy is produced. That leaves one class of observations excluded, namely, those in which entropy is destroyed. Hollinger and Zenzen (1982) concluded:

If we assume that all processes are reversible in the sense of criterion II (reversible in time), then we are forced to notice the exclusion of some time-reversed processes, i.e., irreversibility according to criterion III, and to look for the reason for the exclusions. The absence of those entropy-destroying processes must be due to external causes, maybe historical constraints that keep systems away from certain configurations. Consider, for example, rainfall. Since the rain usually moves downward and rarely upward, we may say that the process of rain is irreversible in the sense of criterion III. It is clear in this case that the irreversibility does not arise from a violation of invariance under time-reversal of the laws of motion but rather from natural constraints on the way raindrops are formed. This is the kind of irreversibility we are especially interested in, and we want to find its origin in the case of phenomena governed by the second law of thermodynamics.

We can examine two additional examples useful in expanding these concepts.

The simple electric circuit (type 3). Consider a system composed of a battery, wires, and a light bulb. Energy flows from the battery through the wires to the light bulb. The bulb lights up and the wires heat up. Energy is lost to the surroundings. When the battery is unhooked, or runs down, the light goes out and the wires cool off. Their molecules return to the same state they occupied before the energy was channeled through them. And yet, the system exhibited irreversible behavior in terms of energy flows even though there was no net change in its structure. There are two main reasons for the irreversible behavior. The first is that the structure of the system allows energy from the battery to flow only into the wires. The second is that the light bulb is not capable of storing energy as well as dissipating it. If we replaced the light bulb with another battery, we would find that we could produce reversible behavior in current flows depending on which of the batteries was better charged. Thus, the irreversible behavior is due to the structural constraints (initial conditions) of the system that exclude some reverse processes.

Convection currents (type 4). If a large pan of water is placed on a small heat source, the water will increase in temperature locally in the area of the heat source, forming a convection current, a visible sign of entropy production, as was the lighted bulb. The convection current forms initially because the slowness of heat diffusion through water creates a temperature asymmetry in the water. Hollinger and

Zenzen noted: "We sometimes think of diffusion as an irreversible process *caused* by collisions. On the contrary. It is an irreversible process, but it is *inhibited* and drastically slowed down by collisions."

If not for this "exclusion" of smooth and rapid diffusion of heat, convection currents could not form because no local temperature asymmetries could occur. Moreover, once a convection current forms, it may suffer two fates. If the heat source is removed, the convection current disappears along with the temperature asymmetry. The system returns to equilibrium, at which point no more entropy is produced. If energy continues to be introduced at a constant rate but not to the boiling point, the convection current will break down into other convection currents as temperature asymmetries are compounded. Note the difference in behavior. In the first instance, return to equilibrium is evidenced by a loss of structural integrity and loss of entropy production (but not entropy destruction). In the second, continued entropy production is evidenced by increasing complexity of the convection currents.

Note also that entropy production is first signaled by the production of a convection current from macroscopically unstructured water, but that subsequently entropy production is signaled by the production of more convection currents, structure from previous structure. And, in each case the energy flow was the same. This suggests that the future manifestation of entropy production in such systems depends on the previous state of the system (in addition to being ultimately determined by the energy flows). Further, we might expect that continuing entropy production would ensure that convection currents would continue to be produced. This continued production is termed by some the *reproducibility* of the system, because the class of allowed structures (in this case, convection currents) is based on the more probable previous configurations and not on the increasingly less probable derivation of the structure de novo from the unstructured original equilibrium. Or, as Hollinger and Zenzen (1982) put it:

> The clue that comes to us from comparison of statistical entropy and the entropy of observed processes is that we should focus our attention on the reproducibility of the observed processes. Because they are reproducible, as signaled by their increasing entropy if by nothing else, they are candidates for statistical descriptions. It follows that their reverses, with decreasing entropy, are not candidates,

not reproducible. The reverses, unknown though they may
be, must be chaotic and unpredictable.

The excluded processes, then, are those which would decrease the
reproducibility of the system by means of energy uptake within the
boundary conditions. In terms of our example, these would be processes that would "unwind" the convection currents. The boundary
conditions can be set as a temperature of not more than 100° C,
because exceeding this temperature makes convection currents inaccessible microstates.

Type 3 and 4 systems are sometimes called nonequilibrium systems. These are *working,* or *entropy-producing,* systems that maintain themselves away from thermal equilibrium. They differ from
type 1 and 2 systems, which perform work only as they move toward
equilibrium. The type 3 and 4 systems differ from each other in that
type 3 systems occupy only a single working state (a *steady state*),
whereas type 4 systems may fluctuate from one steady state to
another through time. Unlike type 1 and 2 systems, nonequilibrium
systems may maintain or even increase their structural complexity
(= *reproducibility* of Hollinger and Zenzen 1982) by producing
entropy.

For the convection current systems, we find that entropy production is characterized by heat flow from the source through the
water. It is also characterized by the formation of convection currents. We know from the structure of the convection currents that
entropy is being produced and that the process of convection current
formation is an irreversible one. We have seen that formation of
convection currents occurs because of an impediment to heat flow
caused by molecular collisions in the water. This would reduce the
amount and rate of entropy production due to heat flows. And yet,
the structural change is also an indication of entropy production.
Unlike type 1 and 2 systems, or any system at equilibrium, where
structural and heat flow changes are isomorphic, structural and heat
flow changes are complementary in nonequilibrium systems.

It is easy to see that biological systems are nonequilibrium entropy
systems, but this is not sufficient for our purposes. We must also
consider the role history might play in determining the manner in
which biological systems attain their nonequilibrium status.

Prigogine, Nicolis, and Babloyantz (1972) presented a summary
equation for entropy production in nonequilibrium systems:

$$dS = d_eS + d_iS, \; d_iS > 0$$

where dS is the change in entropy of the system, d_eS refers to entropy production by energy fluxes from the boundary conditions through the system (heat flow for convection currents), and d_iS refers to entropy production by irreversible processes within the system.

For d_iS to be positive, a system must "export" entropy to its surroundings (Prigogine and Stengers 1984). This entropy becomes the history of the system and forms part of its boundary conditions. Because d_iS is always positive, there is a historical buildup of entropy in the boundaries of any system, and this is responsible for the system's irreversibility—the so-called entropy barrier (Prigogine and Stengers 1984). The boundary conditions are the source of any "selective" influence on a system; so some selective influences may be historically contingent. If the system is relatively random, that is, close to equilibrium, the continual exportation of entropy to the surroundings may produce boundary conditions capable of imposing nonequilibrium status on the system. Note that this view neglects the state of the system itself; hence the restriction that the system must be close to equilibrium. This assumption is justified for any system whose initial conditions play no part in the system's subsequent behavior and status. History is a by-product of system-environment interactions. We believe this is an incomplete view, at least for biological systems.

If energy is taken up by a system, d_eS has a negative sign relative to d_iS. Within the system, the energy can be degraded, or dissipated, in accordance with the second law by being exported to the surroundings (the usual measure of d_iS) or by being associated with the formation of organized structure within the system. If structure is being formed, the absolute value of d_eS is greater than the absolute value of d_iS when d_iS is measured only in terms of entropy exported from the system. Consequently, dS for the system takes on a negative value. This does not mean that the second law is being violated. So long as d_iS, measured strictly in terms of energy lost from the system, is greater than zero, we know that the system is not 100% efficient in using the energy it takes up, something prohibited by the second law. In addition, and more important to evolutionary biologists, under such energetic conditions organized structure always appears in the system and, as Hollinger and Zenzen suggested, this is also an entropic phenomenon.

Kucias (1984) has suggested that the thermodynamic state of $dS < 0$ be termed a *negative entropy state*. He also suggested that the term *negentropy* be used as a "measure of structural arrangement expressed in entropy units," which is "equivalent to the entropy which accompanies the formation of any compound from pure elements." Kucias further emphasized the "impossibility of constructing a system of negative value of entropy." We will avoid using the term *negentropy* for the positive entropy production within a system because of its ambiguity. Thus, we would expect that if d_iS included entropy production associated with structure building, as well as with energy loss from the system, dS would always be greater than or equal to zero, in accordance with universal considerations of the second law. One reason measures of structural entropy are not usually considered by thermodynamicists (aside from the fact that they are not primarily interested in structure) has been the difficulties associated with providing exactly equivalent units. Attempts along those lines involve the fields of statistical mechanics and information theory, which we will discuss next. Thus, irreversible processes always involve energy flows but may also include the formation of convection currents, for example, or any other irreversible structural change. Consider the type 3 system. There is no irreversible change in the structure of the system (the wires and light bulb), but there is irreversible energy flow. Thus, d_iS is manifested by energy flows alone. Alternatively, for the convection current (type 4), we find that d_iS is manifested both by energy flows and structural change. How can we characterize this second aspect of d_iS?

The entropy production associated with structural formation within the system is dissipated into the system itself. It becomes part of the boundary conditions (history) of the past state of the system and simultaneously becomes part of the initial conditions of the subsequent states of the system. Since entropy production is a manifestation of the system's history, this results in at least part of the system's history being incorporated into its causal makeup. Such systems should exhibit both "thermodynamic" and "dynamic" behavior, and the distinction between "initial conditions" and "boundary conditions" will be blurred; initial conditions could well be termed *historical boundary conditions*. The important distinction will be between historical and nonhistorical causal influences. The historical influences will be axiomatic because entropy production dissipated into the system must always affect the system. Nonhistorical influences will be proximal, or contingent, causes because entropy production exported to the surroundings may or may not affect the future of the system.

Since the second law requires that some entropy production be exported to the surroundings, the initial conditions produced by internal entropy dissipation will always be a subset of the total entropy production. Entropy exported to the surroundings will expand the phase space in which the system is evolving, thus moving the equilibrium point farther away; entropy dissipated into the system will cause the system to move toward the equilibrium point. The ability of the system to actually attain the equilibrium point will be constrained to the extent that some of the entropy needed will have been dissipated into the surroundings. Any such system will be in a dynamic nonequilibrium state, spontaneously moving toward an equilibrium point that is redefined by the actions of the system. There will be *emergent* nonequilibrium behavior caused by the initial conditions as well as the possible selective influence of the boundary conditions. We believe that this is a useful view of biological systems. It predicts that their functioning will be entropic, nonequilibrium, and historical at all levels, without sacrificing the selective effects imposed by the environment that have played such an important role in developing biological theory to this point.

Relationship Between Biological Evolution and Thermodynamic Evolution

The living world is not chaotic. In fact, it appears highly ordered, organized, and predictable. The orderliness we observe has two manifestations. First, the living world is composed of integrated working systems and subsystems, ranging from biochemical feedback loops to the patterns of trophic interdependence among species in a community. Some natural systems, like organisms and many species, are discrete and independently functioning units, but they exist within a context of other organisms and species. Virtually all natural processes occur only as part of an integrated whole. Consider, for example, the development of an organism from single fertilized cell to functioning adult. Each of the developmental stages through which the organism passes makes sense to the scientist-observer because it is part of a total integrated program.

The second aspect of natural order is manifested by the observation that the characteristic features of living systems are related to each other in a hierarchical manner. Organisms can be classified, and the great degree of predictability about biological systems stems from the existence of this single natural hierarchy relating all known species.

Despite the extreme amount of orderliness in biological systems, nature is not static. Reproduction and death occur, changing the composition of populations. Development, or ontogeny, occurs, changing the appearance of organisms. Reproduction produces highly faithful but imperfect copies of parents, and development always leads to senescence and death resulting from a breakdown of homeostatic mechanisms. The orderliness and the variability of the natural world have their roots in reproduction, development, and death. All three have one feature in common; they are irreversible processes. This is our clue to the nature of biological evolution. The historical record left by the operation of any irreversible process dealing with discrete units will always be hierarchical, a fact not fully appreciated by biologists until comparatively recently (see Goldschmidt 1952; Hennig, 1966; Riedl 1978). Because we know of no natural hierarchical configurations not produced by an irreversible process, we are logically compelled to accept the proposition that the natural hierarchy of form we see is the result of historical and irreversible processes operating on discrete units.

Three questions must be answered by a more complete theoretical framework for biological evolution. First, why is nature orderly and not chaotic if the living world is constantly in a state of transformation and is incapable of perfect replication? Second, why is the natural world ordered in the way it is? Third, why do particular organisms look the way they do? Biologists have tended to concentrate on the third question. But we contend that before we can confidently answer questions about why certain organisms appear and behave the way they do, we must discover and elaborate a general mechanism for natural order providing a link between the irreversible processes we see transforming organisms and populations today and the hierarchical order produced historically. And the answer to this higher level consideration lies in understanding the physics of irreversible processes as manifested in the singular statement of terrestrial life. Allen (1981) stated: "the word 'evolution' in the physical . . . sciences . . . has traditionally referred to the movement towards thermodynamic equilibrium; the elimination of nonuniformities and the increase of disorder within the system, while in biology . . . it has been associated with increasing complexity, specialization and organization." Evolution in the physical sciences most appropriately refers to those processes in which entropy is produced, but entropy can be produced by systems approaching equilibrium, by those receding from equilibrium, and by those maintaining themselves some distance from equilibrium. We have also seen that "increasing complexity, specialization, and organization" accompany entropy production in systems moving away from, or maintaining themselves away from, thermodynamic equilibrium.

Living systems exchange matter and energy with their environment while maintaining their individuality. A plant absorbs energy from the sun, carbon dioxide from the air, and water and nutrients from the ground. It then uses the sun's energy to convert the water and carbon dioxide into carbohydrates stored in its cells, as well as water and oxygen. Oxygen and heat are dissipated irreversibly into the atmosphere. The plant may utilize the materials and energy taken up to grow in size or complete a developmental step, or it may use those materials for reproduction. A certain amount of material and energy is used to maintain the plant's life functions. Even at rest, a living organism is a working system. Each instance of uptake also produces an irreversible dissipation of matter and energy away from the organism. Even if the uptake and dissipation process changes the appearance of the organism, the organism does not lose its identity, or individuality. That occurs only when the organism dies, when it ceases to be a working system.

The fact that living organisms exchange energy and matter with their surroundings places them in the category of physical systems called *open systems* (von Bertalanffy 1933, 1952). And yet, by virtue of their ability to maintain their identity while undergoing such changes, organisms act as dissipative structures, meaning that they are nonequilibrium systems and are not totally open systems. Rather, they have some definite boundaries, or individual characteristics. Both the persistence and individuality of living organisms are related to their boundary conditions. Mercer (1981) stated:

> It is the source of the boundary conditions that becomes the crux of the matter. Thus a machine is an organized system. Its design specifies the added boundary conditions and their source is the designer. Biological systems are evolved systems. . . . By relating biological and physical systems in a common scheme we arrive at the foundation of a theoretical biology.

Mercer thought, like many others who have considered the problem, that treating organisms as machines was more than an analogy. Taking the place of a human designer was the natural designer, Darwinian natural selection acting to adapt organisms. Taking only this view, however, limits the scope of properties that we envision living systems to exhibit uniquely. Rather than asking how much of the functioning of living organisms can be explained in terms of the "living machine" analogy, we suggest that it may be profitable to ask in what ways do living things differ from machines. By assuming some designing, determinate force, one excludes the possibility of determinate forces that are not goal directed; and so, if any such forces exist, one would fail to recognize them. We have adopted a different approach; we consider the basic components of the boundary conditions themselves in reference only to living things and to see if there are determinate forces that are not goal directed.

Biological systems are clearly a class of nonequilibrium entropy systems. They are, at some level, nonequilibrium thermodynamic systems, but they are also systems whose entropic behavior is not determined (although it is allowed) by energy flows. Many nonequilibrium systems are structurally determined by energy flows through them. Living things are not, and thus should have additional properties not predictable from strictly thermodynamic criteria. These properties will be consistent with the general behavior of nonequilibrium systems but will be defined by the determining capacities (boundary and initial conditions) at the new functional levels.

Boundary and initial conditions for organisms. It is a truism that living organisms must take up, utilize, and dissipate energy or they will die. It is also true that no living system can survive whose minimum needs exceed available energy supplies. From this, one might then assume that energy flowing through the surrounding environment creates the boundary conditions for organisms. But we would assert that the flow of energy cannot explain the structure of living organisms. Energy is modified or differentially utilized by organisms, and that modification or use is determined by properties intrinsic to the organisms. These intrinsic properties characterize an organism. In this regard, we agree with recent workers such as Wicken (1979) that energy flow is open and essentially unlimited as far as living organisms are concerned, that is, more energy reaches the earth from the sun than is ever used by living organisms. Energy flows do not provide an explanation for why there are organisms, why organisms vary, or why there are different species.

If energy flows were the determining factors for biological organization, then we would expect to see changes in energy flows causing changes in that biological organization. On a gross level, we would expect to see structural change lagging behind ecological change (form following function). However, changes in biological organization have their bases in mutation and there is no link between mutation and energy flow that is analogous to the role of energy flow in the organization of nonliving physical systems. It is an organism's intrinsic properties that determine how energy will flow, not the opposite. Consider a hummingbird in a pasture. The sun is shining, there are many insects and other small animals in abundance, there are many plants, and there may even be radioactive rocks in that pasture. In short, there is an abundance of energy. However, in the absence of certain kinds of flowers, that hummingbird will die. Something other than the flow of energy determines the individual characteristics of living organisms. If the flow of energy were deterministic for biological systems, it would be impossible for anything living to starve to death. We conclude that the flow of energy is a relatively stochastic factor for biological systems.

Individual organisms are initially bounded, or partially closed, from the standpoint of "epigenetic information," that is, the sum total of genetic and cytoplasmic information specifying the structure of these organisms. We are intrinsically limited by our ontogenetic program, specified by our genes and our cytoplasmic organization. We do not receive genetic instructions from an outside source subsequent to conception. Thus, the genetic program and cytogenetic organization comprise at least part of the intrinsic constraints, or initial conditions, which make organisms partially closed systems.

We suggest that living organisms are physical systems with genetically and epigenetically determined individual characteristics, which utilize energy that is flowing through the environment in a relatively stochastic manner. A general characterization of dissipative structures is that they are physical systems in which at least one stochastic and one determinate factor interact. The interaction of finite epigenetic information, determined by egg and sperm, with a sufficient, stochastic flow of energy establishes the stochastic-determinate dynamics that permits an organism to survive. The organism's epigenetic information allows certain forms of energy to be utilized for homeostasis (maintenance), ontogeny (growth and differentiation), or homeorhesis[1] (reproduction). The energy taken up is used to produce metabolic waste products, heat, biochemical changes, and complex structures. Each of these processes involves a series of metabolic pathways for converting matter and dissipating energy. As properties of a single cohesive ontogenetic sequence,

---

1. Literally, same-flow (see Waddington 1977). We know reproduction is homeorhetic because we observe more replication than mutation and assortive mating.

these pathways for dissipation are causally linked. In chapter 3, we will discuss predictions deducible from the hypothesis that this ontogenetic linkage produces *alternate dissipative pathways* related in such a manner that changes in one result in compensatory changes in one or more of the others.

Boundaries of survival. If developmental programs were infinitely long, ontogeny would continue indefinitely. And, if homeostatic mechanisms were perfect, an organism could maintain itself forever, even after ontogenesis was completed. However, we do not know of any cases of infinite ontogenies or of perfect homeostatic mechanisms in organisms. Indeed, some cell lineages have finite "life spans" that can be measured in generations. Thus, no single organism can escape death, or thermodynamic equilibrium. However, a genealogical lineage of organisms may escape thermodynamic equilibrium through reproduction.

Any living organism that dissipates some of the energy it takes up into reproduction may thus contribute to an escape from death. Its lineage, or *species,*[2] maintains itself *at any given time* through the homeostatic mechanisms of its living individual organisms. The species develops and maintains itself *through time* by means of high-fidelity (but not necessarily perfect) reproduction. Species are thus energy-transforming systems and can be considered thermodynamic systems, with energy retained in the system as structure through reproduction. Thus, the "ontogeny" of an individual species is its sequence of reproductive events. The "ontogeny" of a species is open ended, or potentially unlimited, because we do not know of any limits on the number of reproductive generations for a lineage (the idea that species, like organisms, succumb to "old age" was laid to rest by Simpson, 1944).

Every organism has a birthday. Organisms maintain their identities through time, despite structural changes accompanying ontogeny. This spatiotemporal continuity is achieved through the interaction of genetic and epigenetic information, organized in an ontogenetic program, and energy flowing through the environment. The interaction of the deterministic ontogenetic program and stochastic energy flows produces living dissipative structures; dissipation is achieved through a variety of pathways specifying homeostatic developmental and reproductive activities.

Individual living systems are dissipative structures. They have a single origin in space and time, exhibit spatiotemporal continuity

2. See Wiley (1978, 1981c) for a discussion of species as lineages.

(maintaining their individuality through time and changes), and possess intrinsic boundary conditions, or identifying traits. Reproducing individual organisms produce parts of the kind of individual living system that can escape thermodynamic equilibrium—the individual species. Species are individual living systems so long as they have a single origin in space and time, exhibit spatiotemporal continuity, have identifying traits, and are dissipative structures. They are dissipative structures so long as they retain individual identity (their parts, or individual organisms are more "cohesive" (defined later in this chapter) with each other than with parts of other such systems) and so long as their parts exchange matter and energy with the surrounding environment.

Metaphysics of individuality. The characteristics we ascribe to particular populations and species and our view of the nature of species as entities will profoundly affect the way we deal with them and what processes, if any, we think they might participate in. Ghiselin (1966, 1974, 1980, 1981) has developed the idea that particular species such as *Homo sapiens* are, ontologically, individuals rather than natural kinds. Natural kinds constitute universal classes; the problem with universal classes is that they are eternal and immutable (Ghiselin 1966, 1974, 1980, 1981; Hull 1976, 1980, 1981, 1983; Wiley 1980a, 1981a), and attempts to render particular species as exceptions (i.e., natural kinds that can evolve) has done nothing to clear up the matter (Greene 1978; see comments by Hull 1983). The problem with classes is that they do not participate in natural processes nor are they the result of particular and unique histories. This does not mean that natural classes are not important. Hydrogen is a natural class, and it is important because its definition allows physicists to deduce the behavior of the individuals that fill the class. "Evolutionary species" is a natural class and is filled with particular species who have the characteristics that define species. But neither "hydrogen" nor "evolutionary species" participate in natural processes. In contrast, hydrogen molecules and *Homo sapiens* participate in all sorts of natural processes. A hydrogen molecule is the thing which reacts. *Homo sapiens* and other particular species are the things which evolve. Such individuals differ from classes by having particular origins in time and space. They may disappear or change into other things.

The metaphysical view of species as individuals is being widely adopted by biologists (Hennig, 1966, did so independently of Ghiselin; Griffiths 1974; Mayr 1976, 1978; Wiley 1978, 1980a, 1981a; Patterson 1978; Eldredge and Cracraft 1980), because it provides a cogent metaphysics to appreciate species as evolutionary units. A temporal sequence of transforming individuals implies an inherent time asymmetry in the process of transformation. If we view biological evolution as a historically constrained process, we must reject the metaphysics of immutable classes. This means we must reject theories in which ecological constructs ("niches," "adaptive zones") serve as analogues of quantum states and the biological entities occupying them are contingencies determined by the energy flowing from the sun to the earth. Rather, we must view those ecological constructs as contingent energy flow pathways determined by the constraints on energy dissipation inherent in the historical order of the "individuals" undergoing ontogeny, reproduction, population change, speciation, or community evolution. This is not to say that the ecological components are in any sense unreal or unimportant (see chapters 4 and 6).

We may ask an additional question. Are all species strictly individuals, that is, do all species show both continuity and cohesion? The answer, of course, is no. Asexual species show continuity but no cohesion. Species composed of allopatric subpopulations show no overall cohesion. Such species are *historical groups* (Wiley 1980a), and the units of actual change are individuals and subpopulations, respectively. Being an ontological individual but not an actual individual does not mean that such taxa are classes; they are still restricted to particular origins, and their parts may transform. Populations of sexually reproducing species may be parts of actual individuals or parts of historical groups. If a population is part of an actual individual, it shares genes with other parts and thus is an open information and cohesion system. Changes that affect one part can potentially affect all parts. If a population is part of a historical group termed *a species,* then it, itself, is a closed information and cohesion system with respect to the rest of the species. Changes that affect it cannot affect the other populations. It is an actual evolutionary unit in its own right.

Some taxa with more than one species are also historical groups (Wiley 1980a). Natural higher taxa (the monophyletic groups of Hennig 1966) are no more classes than are species. However, they have the classlike characteristic of being composed of independent individuals capable of evolving (or changing) independently of other included individuals. Their "naturalness" depends on whether they are complete historical units (monophyletic groups). Other higher taxa, nonmonophyletic ones, are classes (Ghiselin 1980) and thus eternal and immutable as well as being inconsistent with the phylogeny of their constituent members (Wiley 1981b). They are "un-natural" and have no place in the scheme of things from the evolutionary point of view. This point is fairly important to tax-onomy in general because it means that such groups as Reptilia and Protista must be abandoned if taxonomies are to be evolutionary constructs.

Boundary and initial conditions for species. If species are dissi-pative structures, they must exhibit boundary conditions. Species cannot be completely closed, or bounded, systems, or they would be unable to exchange matter and energy with the outside world and life would be impossible. On the other hand, if species had no bound-aries, there would only be one of "them." "They" would not exhibit dissipative behavior and would not be able to maintain themselves in a viable state (i.e., a nonequilibrium thermodynamic state). We may recognize that species are partly closed systems simply from the observations that there are more than one species and that rep-lication rates are higher than mutation rates. They are bounded, or partially closed, because different species either do not share, or share only to a limited degree, reproductive linkages and information specifying dissipative pathways. They are open systems to the extent that as long as they remain minimally cohesive (i.e., reproduction continues to occur), they may continue to process energy and, as we will show, may change their information systems without losing their individuality. In this regard, populations within sexually re-producing species that exhibit gene flow are open systems in terms of information and cohesion (reproductive linkages), exchanging in-formation via cohesion bonds. "Island" populations (i.e., those without gene flow) and asexual clones are closed systems, as are reproductively isolated species, because they do not exchange in-formation and lack cohesion.

A less metaphysical and more pragmatic demonstration of the closure of certain biological systems is elucidation of any self-organizing or anamorphic (complexity-generating) properties of those systems. All physical systems are transformational so long as they are not in thermodynamic equilibrium. Dissipative structures, or individualized open systems, use energy flow to maintain themselves in a dynamic, nonequilibrium state. Species, as living systems partly closed in terms of information and cohesion but open in terms of energy, should exhibit two kinds of processes: (1) transformational and cyclic processes such as the Krebs cycle or feedback loops and (2) transformational and noncyclic processes such as ontogenesis and phylogenesis. If a system is partly closed, transformational, and noncyclic, then each transformation will be historically unique to the individual in which it occurs and will affect only part of the total system (or superindividual) to which the individual belongs. Summation of a series of transformations will produce a hierarchical sequence of historical change.

The evidence that evolution is transformational and noncyclic rather than cyclic comes from two sources. First, the parts of species, organisms, exhibit ontogenies that are transformational and noncyclic. Second, our empirical data base (see chapter 5) indicates that species are hierarchically related to each other and that each clade (group of species hierarchically related) is characterized by the presence of some characters that indicate its historical uniqueness (transformed parts of the system), or *apomorphies* (Hennig 1950a), while at the same time it has other characters that show its roots (nontransformed part of system), or *plesiomorphies*. If evolution were cyclic, we would expect no hierarchy of taxa and characters or we would expect a different hierarchy for every set of characters we analyzed.

That evolution produces a hierarchy has long been understood (see Darwin 1859; Hennig 1966), if not always appreciated. But why should it? Because, as some authors (e.g., Goldschmidt 1952; Hennig, 1966; Riedl 1978) have correctly pointed out, if evolution is an irreversible process working on discrete units, then a hierarchy is the result. Our theory suggests that evolution is a phenomenon involving systems (species) far from equilibrium. The hierarchy results from speciation, which we will try to show exhibits entropic dynamics *analogous* to "ordering through fluctuations" (Prigogine, Nicolis, and Babloyantz 1972). It is important to understand that this is an analogy. Ordering through fluctuations in strictly thermodynamic systems is a direct by-product of energy flows. Ordering through fluctuations in evolution is a direct by-product of information and cohesion changes and not, we submit, energy flows. Such a phenomenon, representing a general law, would produce, as a universal statement, a historically unique hierarchy. However, the theory would not predict the singular statement of the unique phylogeny produced by organic evolution on this planet. Questions concerning causality of particular episodes in the hierarchy depend on the parameters covered by the next lowest order theories and the research programs or predictions deduced from them, which serve as bridge principles (Hempel 1966) between empirical observations and the highest level theory.

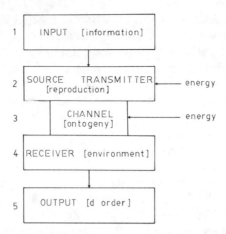

Fig. 2.1. Schematic flow diagram depicting relationship between the flow of biological information and the flow of energy from the environment.

The Nature of the Initial Conditions

Information

Information, in a primitive sense, may be defined as anything transmitted from a source, through a channel, to a receiver (Gatlin 1972) (fig. 2.1). This use of information differs from "information" as used in communications theory, where information is essentially defined as the ability to transmit the kind of information with which we are concerned. Information, like energy or gravity, is a capacity. Energy is the capacity to do work. Information is the capacity to execute an ontogenetic program and/or maintain homeostasis (provide a continuous energy and matter exchange between the organism and the environment). Waddington (1977) referred to this as instruction rather than information. We may think of this information as being of two basic kinds, canalized (or regulatory) and noncanalized (or structural). Canalized information is that information responsible for the sequence of ontogenetic events the organism goes through during its development and any genetically based behavioral traits it may exhibit. Included is information from regulatory genes, cytoplasmic organization, cell interactions (e.g., morphogenetic movements), and inducers (for recent discussions of such phenomena see Løvtrup 1981; Horder 1981). Noncanalized information is that information derived from structural gene loci. Noncanalized information is used to produce structural products (proteins, enzymes, etc.), which in turn are used as building blocks for realizing the ontogenetic transformation of the individual.

Obviously, not all information, canalized or not, can be expressed all of the time. In fact, some information is never expressed, either because it is blocked or is never induced. We distinguish between these two types of information using terms derived from information theory. *Stored information* (or expressed information) is unambiguous information; for an organism, this is information actually expressed at one or several times during the life of the individual. Stored information acts to order and/or integrate the organism's functioning. *Potential information* (or unexpressed information) is ambiguous information; for an organism, this is information that is present but not expressed. A simple example of potential information in an organism would be a recessive allele in a heterozygous individual. Another example would be an alternate development pathway that could be expressed given the right signal, but is not. Given these distinctions, we may now characterize the different sorts of information associated with living organisms (we do not wish to imply that these classes are mutually exclusive or are the only possible classes).

1. *Canalized stored information.* This could also be called expressed regulatory information, because it refers to the expressed sequence of ontogenetic and genetically determined attributes an organism displays during its lifetime. We have termed the product of any particular individual canalized stored information system the *epiphenotype* (Wiley and Brooks 1982). This corresponds to the sequence of *semaphoronts* of Hennig (1966).

2. *Canalized potential information.* This term, for which unexpressed regulatory information could be substituted, comprises alternate ontogenetic pathways or alternate behavioral traits not expressed in the individual because they are blocked by the expression of other information or have not received the appropriate signal for expression.

3. *Noncanalized stored information.* Also termed expressed structural information, this aspect of an organism's total information system includes homozygous alleles, heterozygous co-dominant alleles, or dominant alleles at heterozygous loci coding for structural gene products.

4. *Noncanalized potential information.* This component could be called unexpressed structural information. It includes recessive alleles in the heterozygous condition coding for structural gene products.

Organisms may also have *new information* that acts as canalized or noncanalized potential information in the gametes of the individual and may or may not be expressed in subsequent generations. This new information might be a point mutation, duplication, chromosome mutation, or any number of such things. We will consider mechanisms for the production of new information in chapter 3. The extent to which new information is actually new, that is, a true evolutionary novelty, will depend on whether the information is the product of a historically unique event or a recurrent event.

We have suggested that species are information systems. As such, species should have information classes analogous to those of individual organisms. The canalized stored information of a species is the set of ontogenetic steps executed by all viable and fertile members of the species. Some members of a species may lack some of these steps, but they have lost the evolutionary game before they begin. So, the continuity of the species depends on the members of a subset of its parts—those individual organisms that have the species canalized stored information system. The morphological similarity among members of a species is explained by the fact that those organisms we can examine have correctly executed the canalized stored information system of the species, at least to the point in ontogeny where we observe them. Variations in morphology (or any other genetically determined processes) are due to *species-level* canalized potential information, which may be defined as variations due to differences in the canalized stored information systems of different individual organisms. Thus, at the species level the following classes obtain: (1) Canalized stored information pertains to that information in all viable and fertile individuals that does not vary among members of a species. (2) Canalized potential information refers to the canalized stored information in individual organisms that varies among members of a species.

Species also have noncanalized information. Noncanalized stored information is represented by fixed gene loci coding for structural products, whereas noncanalized potential information is represented by loci with two or more alleles that code for structural products.

Individual organisms of a species resemble each other because they share a unique canalized stored information system inherited from their parents. Maintenance of the identity and discreteness of a species, the species' homeostasis, requires high-fidelity replication to transmit the species' canalized stored information system from generation to generation. Species differ from each other because they have, to a greater or lesser extent, different information systems. Differences between the noncanalized information of two species may be measured in a variety of ways, including DNA and amino acid sequencing as well as electrophoresis. Such differences can be valuable in discovering genealogical relationships and may be strongly correlated with speciation. However, changes in noncanalized information are not sufficient to explain the origins of most species. Rather, the mechanical explanation for organismic diversity lies with modification of the canalized information system, including "regulatory genes" in particular (Britten and Davidson 1969; Templeton 1981; Kauffman 1983), epigenetic modification in general (Goldschmidt 1940; Waddington 1957; Løvtrup 1974, 1981; Horder 1981; Rachootin and Thomson 1981), and chromosomal changes (Stebbins 1971; White 1973, 1978; Bush et al. 1977). We may think of species as individualized, evolved, canalized stored information systems. The reason that *Homo sapiens* looks and acts differently from *Gorilla gorilla* is that the two have different canalized systems. However, these systems are not completely different. Barring convergence, two species will share a common canalized information system to the extent that they share common ancestors, and they will differ to the extent that their common canalized information system has been modified after their common ancestor speciated. Thus, we suggest that speciation must ultimately be explained in terms of converting canalized potential information contained in an ancestral species into canalized stored information in its descendants.

A species that has little or no canalized potential information and thus little structural variation may be completely described with relatively few bits of information. No matter how many components (member organisms) such a species contained, it would only take a short description to individualize and characterize it. Such a species would be considered very ordered and simple in information theoretical terminology (Chaitin 1975; Wicken 1979 and references contained therein). A species showing no structural or behavioral variation, that is, having no canalized potential information, if such a thing could exist, could be described by a single member of the species. Each individual would be analogous to a component of a monotone, and the species would be at a state of minimum information entropy (Gatlin 1972).

A species with canalized potential information requires more information to describe it. A species composed of two different canalized information systems would be relatively more complex and disordered than a species having a single canalized information system. Complexity due to potential information has been termed *het-*

erogeneity by Wicken (1979; see also Ho and Saunders 1984), and we use the term here as it applies to populations of whole organisms. Complexity due to heterogeneity should not be confused with complexity due to the size of one information system compared to another. One information system may be more complex than another simply because it specifies more ontogenetic steps, as shown below:

Epiphenotype I: a $\longrightarrow$ b $\longrightarrow$ c
Epiphenotype II: a $\longrightarrow$ b $\longrightarrow$ c $\longrightarrow$ d $\longrightarrow$ e

Epiphenotype II is more complex due to an increase in the size of its canalized information system. If a species contained individuals with both epiphenotypes, it would be complex due to heterogeneity (two epiphenotypes) and complex due to size (the ontogenetic programs are of different lengths). However, complexity can be produced by substitution, as shown below:

Epiphenotype I: a $\longrightarrow$ b $\longrightarrow$ c $\longrightarrow$ d
Epiphenotype III: a $\longrightarrow$ b $\longrightarrow$ c $\longrightarrow$ f

A species containing epiphenotypes I and III would show as much heterogeneous complexity as a species containing epiphenotypes I and II, in spite of the fact that the species containing epiphenotype II has the potential to produce a descendant species more complex in size than its ancestor.

Cohesion

An organism maintains its identity during ontogenesis partly because it is a closed system in terms of information content. However, equally important is the recognition that the information is organized in such a manner that the organism exhibits spatiotemporal continuity during ontogeny and that each stage maintains the viability of the organism. This results in ontogenetic programs being causal sequences rather than collections of forms individually produced (see chapter 3 for a fuller discussion of this point). An organism's cohesion is the unity and organization of its information, manifested by such phenomena as cell adhesion and physiological integration.

Reproductive interactions between individuals within a species will determine that species' cohesion. The ontogeny of a species is provided by reproductive continuity, which is responsible for giving an individual lineage the appearance of a causal sequence. We may view the actual reproductive interactions within a species as a linkage pattern or network. Given a particular species, we may measure cohesion in two ways: (1) as a function of the probability that individuals representing two different epiphenotypes can and will successfully mate and (2) as a function of the probability that individuals of any one deme will mate with individuals of other demes. If the number of potential linkages is high in both cases, then the linkage network approaches panmixis, and the species is highly cohesive. All epiphenotypes mate with each other equally often and equally successfully both within and between demes. Species that show a high degree of cohesion may be said to be highly organized and relatively simple. The less panmictic a species is, the more disorganized and more complex the species becomes. One source of complexity and disorganization is a manifestation of the presence of two or more epiphenotypes. If, for example, two epiphenotypes tend to show assortive mating, forming two rarely interconnected linkage networks, then the species shows a degree of disorganization and disorder directly related to information complexity. On the other hand, if the same species exhibits panmixis, the species remains more organized and ordered than could be predicted simply by assessing information complexity.

A species may also exhibit some disorganization and thus be more complex simply because it is spread over enough geographical space to make panmixis physically impossible. Any species occupying a geographical area larger than the range of a member of that species (as defined by Endler 1977) exhibits a certain amount of disorganization due not to informational complexity (although this will augment the complexity) but to the impossibility of maintaining a panmictic linkage network.

Asexual species, as noncohesive entities, exhibit disorganization. Sexually reproducing species composed of island populations with no gene flow are also disorganized because there is no cohesion between demes. Disorganized systems have the interesting evolutionary capacity of being composed of parts that may evolve independently of each other.

Measuring the Entropic Behavior of Living Systems

The functional attributes, that is, the irreversible processes, exhibited by living systems that are components of biological evolution are (1) *homeostasis* and *metabolism*, (2) growth and development, or *ontogeny*, (3) *reproduction* and *population differentiation*, (4) *speciation*, or lineage splitting, and (5) assemblage into *ecological associations*. If we are to investigate these as entropic processes, we must be able to measure entropy production associated with each.

If living systems are nonequilibrium systems of the type we have
proposed, there are two components of d_iS that can be used to
discern entropic behavior, energy flows and structural change. But
what units of measure are appropriate? For example, can we equate
genetic heterogeneity with energy in motion?

This is a point at which the "physical" and the "biological" view
of biological systems may diverge. To most physicists the measure
of entropy production, as an abstract quantity or a rate function,
derives only from energy flows, that is, thermodynamic entropy.
Energy states "evolve" in most systems studied. And yet, for most
biologists, it is structure and not energy that evolves. For strictly
thermodynamic systems, the energy flows determine the boundary
conditions, and yet biologists observe living systems changing their
boundary conditions in the presence of stochastic energy flows. And
finally, while it is readily seen that at least some aspects of homeostasis and metabolism, as well as growth and development and ecological association, are accompanied by changing energy flows, it is
not so clear that reproduction or population differentiation and speciation are determined by changes in energy flows. Presumably, this
is what led Denbigh (1975) to assert that most processes relevant to
biological evolution have nothing to do with thermodynamics.

We will attempt to show in subsequent chapters that for processes
1 and 2, both energy flows and structural changes are important
components, and thus those processes can be readily measured in
terms of either. We will also attempt to show that for processes 3,
4, and 5, energy flows are not proximally determinate, and thus the
entropic behavior of those processes are more readily measured in
terms of structural change.

Measures of Thermodynamic Entropy

Two observations pertain to the question of energy flows and
biological evolution. The first is that biological processes, at least
on this planet, occur within a very restricted absolute temperature
range, indicating a significant degree of temperature constraint on
evolution. The second is that many morphogenetic and other developmental pathways are highly temperature sensitive. This limits
the degree to which evolutionary changes can be effected by changes
in the magnitude of external energy flows.

Prigogine and Wiame (1946; see also Prigogine 1947, 1967) provided the first attempt to extend the main principles of nonequilibrium thermodynamics to development, growth, and aging with respect

to energy flows. They considered the entire life of an organism to represent a linear sequence of continuous approach to a final steady state (the "adult" stage), accomplished by a decrease of the specific rate of entropy production (dS) of a specific dissipation function, ψ. Zotin and Zotina (1978) asserted that ψ can be approximately equated with the intensity of heat production (g) or of respiration (g_{0_2}) of organisms, such that

$$\psi = g = g_{0_2}$$

thus reducing the Prigogine-Wiame postulate to the assertion that development, growth, and aging are accompanied by a continuous decrease in the intensity of heat production or respiration. Expanding on that theme, Zotin suggested that living organisms might not develop in a linear, that is, nonfluctuating, manner to a final steady state, but rather might proceed through a series of steady states during ontogeny. Furthermore, the dynamics of this system might be expressed as

$$\psi = \psi_{(o)} + \psi_{(i)} + \psi_{(a)} + \psi_{(k)}$$

where $\psi_{(o)}$ is the specific dissipation function for basal metabolism; $\psi_{(i)}$ is the specific dissipation function for energy flows associated with inducible phenomena, or those internal or external factors contributing to a temporary deviation from or approach to an intermediate steady state (in other words, homeostatic mechanisms); $\psi_{(a)}$ is the specific dissipation function for so-called *inducible adaptive* processes, or the irreversible fluctuations from one steady state to another during ontogeny; and $\psi_{(k)}$ is the specific dissipation function associated with so-called *constitutive* processes, or structure building.

The most promising aspect of this work, which we will discuss in more detail in chapter 3, is that it shows that macroscopic processes associated with ontogeny, growth, and aging of individual organisms, including formation of structures, can potentially be measured in terms of energy flows through the system. If this is true, then the inability to measure the thermodynamic behavior of living systems at the level of reproduction, population differentiation, and speciation must be attributable to technical difficulties rather than to the latter three processes being qualitatively different kinds of physical processes.

Erneux and Hiernaux (1979) and Lacalli and Harrison (1979) have shown deterministic aspects of morphogenesis to be deducible, under certain initial conditions, from the kinetics of reaction-diffusion equations, in agreement with the conditions for fluctuating nonequilibrium systems summarized by Allen (1981) (see previous section).

In more directly biological terms, new structures made by the organism using energy taken up at one time may alter the organism's subsequent energy uptake capabilities. This often occurs during development. For example, larval frogs eat plant material, deriving energy from it and then dissipating some of the energy into development and metamorphosis into an adult frog, which is a carnivore. In this case, both tadpole and adult may be considered as occupying different steady states. The structure of each stage is represented by deterministic constitutive processes characteristic of dissipative structures between fluctuations. In fact, all of ontogeny may be viewed as a series of fluctuations from one steady state to another (Ho and Saunders 1979; Saunders and Ho 1981). Once all ontogenetic commands have been executed, the organism's homeostatic mechanisms maintain its viability. This corroborates Zotin and Zotina's (1978) expansion of the Prigogine-Wiame model. None of this, however, addresses the question of the source of energy uptake and structure formation instructions.

Measures of Statistical Entropy

Measures of statistical entropy trace their origins to Boltzmann (1877), who was seeking a way to characterize the distribution of energy in a system into various states that maximized the state probability of the system. Since thermodynamic equilibrium was the universal endpoint ("attractor") of all systems known at that time, equilibrium should be the "most probable" state. Boltzmann found his statistical measure in formulas developed by de Moivre (1756) to explain the odds in various games of chance. The approach is based on the notion that any given system must pass through all possible microstates before returning to original ones. At equilibrium, the system has passed through all possible microstates and is free to access all microstates randomly. A statistical system may be viewed as an *ensemble* of microstates. Gibbs (1902) developed ensemble theory as a major part of statistical mechanics. The function of ensemble theory was to characterize systems for which exact initial conditions were not known in order to follow the system's behavior through time.

A system, or ensemble, that is not at equilibrium will exhibit a statistical entropy that is lower than the maximum possible (achieved at equilibrium). Shannon (in Shannon and Weaver 1949) first linked the statistical entropy of a system and its *information*. Jaynes (1957a, 1957b), Yockey (1958), Tribus (1961, 1983), Evans (1966), Gatlin (1972), Elias (1983), and Montrol (1983) are among those who have supported the connection between statistical mechanics and information theory. This has permitted a formulation of the second law that is highly appropriate for our purposes (Layzer 1977): In a natural macroscopic system macroscopic information is present and microscopic information is absent.

Layzer noted that classical thinking in physics asserted that only at thermodynamic equilibrium will microscopic information be truly absent. Otherwise, microscopic information is present but simply lacking in our description of the system. In principle, one could obtain such information by expending free energy. Layzer suggested that this rationalization was a weak one for a physical principle as important as the second law. He suggested that microscopic information was truly lacking in macroscopic systems and offered the following formalization (Layzer 1977): "A complete statistical description of the universe defines no preferred position or direction in space; it is invariant under spatial translation and rotation." This has two outcomes, each of which is important to our considerations (Layzer 1977). First, given a single realization of a complete statistical description, one can in principle estimate any parameter of the system with arbitrary precision. We will take a "single realization" in the first property to refer to a single history for any macroscopic system. Furthermore, since the first property derives its justification from the "law of large numbers" (Layzer 1977), we will assume that, given description of enough components at any functional level in a hierarchically structured system, any macroscopic behavior can be estimated with "arbitrary precision." Layzer (1977) stated: "Biological organization is hierarchical; so is information. . . . Thus it allows one to deal directly with the distinctively biological aspects of, say, a macromolecular system while ignoring the biologically less interesting physico-chemical properties."

The second property provides strong support for the notion that there is no microscopic information. It states (Layzer 1977): "Two realizations that are macroscopically (i.e., statistically) indistinguishable are also microscopically indistinguishable. That is, the statistical description is *complete*." If two realizations that cannot be distinguished macroscopically also cannot be distinguished microscopically, the microscopic aspects must not contain information. By extension, our ability to distinguish different realizations as being different must stem solely from macroscopic information. That macroscopic information is defined as $I = H_{max} - H$, where H_{max} is defined as the maximum entropy given the (statistical) constraints and H is a coarse-grained entropy derived from observed or inferred state variables (discussed next). The value of I may decrease, if the system's behavior brings it closer to equilibrium (i.e., H increases until $H = H_{max}$, hence $I = 0$). Thus, equilibrium systems are said to be information-destroying systems. Alternatively, I may remain constant or increase if auxiliary conditions (initial or boundary) are capable of maintaining the system at some distance from equilibrium. In other words, nonequilibrium systems are inherently information-producing (or conserving). The question of relevance for evolutionary biology is the manner in which the information is produced, which parallels the question of how living systems attain their nonequilibrium status, which we have already discussed (see also Brooks and O'Grady, submitted for publication).

Macroscopic structural change in entropy systems is generally measured by assessing changes in the number of microstates accessible to the system at a given time. The assessments are given in terms of probability statements about the behavior of the system, that is, the statistical entropy. The field that addresses these problems is *statistical mechanics*. To show the connection between statistical mechanics and the measures of structural entropic change used in this book, we will employ the simple model presented by Hollinger and Zenzen (1982).

The statistical entropy of a simple system may be given as:

$$S = k \ln \Omega$$

where Ω is the set of accessible microstates for the system (the "macrostate"). The term Ω may be expressed in a probabilistic sense by W, the set of microstates accessible to the most probable distribution. This equation would thus be transformed to

$$S = -k \sum P_i \ln P_i$$

where $P_i = \dfrac{1}{W} \left(\text{or } \dfrac{1}{\Omega} \right)$. ($H$ and S are alternative symbols.)

At thermodynamic equilibrium, the number of microstates accessible to the system is maximal. However, none of those microstates is peculiarly associated with that system; that is, all microstates are equally accessible to all possible macrostates. Therefore, the behavior of the system is *periodic*, that is, one could say that during a particular time interval, all microstates will be accessed or, conversely, that each microstate will be accessed in a determinate manner periodically, and the periodicity can be predicted statistically. Any fluctuations in the macrostate composition, therefore, do not represent departures from or approaches to equilibrium because they are embodied in the probability statement of equilibrium. Thus, no entropy is being produced at equilibrium.

Systems approaching thermodynamic equilibrium exhibit behavior characteristic of the classical view of entropic behavior. The system becomes more periodic in its behavior as it accesses more microstates and as its microstates are accessed by an increasing number of other macrostates. Because it is accessing more microstates, the system is becoming more "probable," that is, more determinate in its statistical behavior. The system is also losing its individuality as more of its microstates are accessed by other macrostates. Both increasing periodicity and decreasing individuality are associated with decreasing reproducibility. If its microstates are periodically being accessed by other macrostates, then the system as a discrete entity can exist only periodically itself. And the approach to equilibrium is characterized by increasing periodicity. Thus, we can see that thermodynamic systems approaching equilibrium are characterized by decreasing structural order and organization, increasing probability, and increasing periodicity. All of this is manifested by statistical entropy production.

One might logically conclude that departures from thermodynamic equilibrium are the inverses of approaches to equilibrium. Therefore, they must be characterized by increasing structural order and organization, decreasing periodicity, decreasing probability, and destruction of entropy, that is, negentropic behavior. Such a conclusion would be only partly correct, because departures from equilibrium, while they are characterized by increasing structural order and decreasing periodicity, are also characterized by increasing probability and positive entropy production. This can be shown most easily by considering the concept of statistical entropy again.

The production of entropy in a statistical sense is associated with an increase in the number of microstates accessible to (accessed by, at least periodically) a macrostate. Any macrostate that moves away from equilibrium has a historically determined set of microstates. As this system moves farther away from equilibrium, it will produce entropy if it accesses new microstates while retaining its historical access to the initial microstates. In such a case, we would find an increasing number of microstates accessed, and hence the macrostate would be an increasingly *probable* system. However, because the system would retain its historical access to initial microstates, it would be accessing an increasing number of microstates in an aperiodic, increasingly reproducible, manner. Entropy production in systems moving away from equilibrium involves increasing organization and decreasing periodicity.

The most probable state for a system approaching equilibrium is, as we indicated earlier, the one in which the maximum number of microstates is accessible to the macrostate and all microstates are equally accessible to all other macrostates. Systems approaching equilibrium thus become progressively more probable in this sense. For a system leaving equilibrium, there is a single macrostate which is the ensemble of microstates defined by the historically determined microstates plus all other microstates accessed by the system since it left equilibrium. Each state is thus more probable than its predecessor and is the most probable configuration at that time. Consequently, the next state, which will involve accessing new microstates, will be even more probable in this sense even though they will be more complex ensembles. Systems moving away from equilibrium, or nonequilibrium systems, thus become "more probable" as they evolve into more complex structures (see also Wicken 1979). And yet, they exist in the most probable state at any point in time, because they embody their own history. This is what allows nonequilibrium systems to evolve without disobeying the second law of thermodynamics. As a simple example, consider the oxygen-binding affinities of hemoglobin, which diverges markedly from the standard Michaelis-Menton binding affinity curve. The Michaelis-Menton curve asserts that most binding sites will be filled during the earliest part of the reaction. Hemoglobin is a complex molecule with four binding sites for oxygen. Given no oxygen bound to the molecule, there is a certain probability associated with the acquisition of one oxygen and a vanishingly small probability of binding four oxygens at once. However, once a single oxygen is bound, the probability of binding another increases markedly and so on until all four binding sites are filled. The final state was highly improbable a priori but became more probable as the system changed.

The only way in which a system could behave "negentropically" would be if it moved away from equilibrium, lost its historical microstates (its *historical entropy sensu* Hollinger and Zenzen 1982) and failed to access any new microstates (the *dynamic entropy* of Hollinger and Zenzen). We have seen that any periodic move away from, or approach to, equilibrium involves no net entropy production, because no new microstates are ever accessed and no historical microstates are lost. We have also seen that aperiodic, reproducible macroscopic behavior involves, by necessity, the accessing of new microstates. Or, as Hollinger and Zenzen (1982) stated: "The *historical* entropy must remain constant in time, even in more complex models, because it measures only the microstates which have evolved from the initially accessible microstates. The *dynamic* entropy must grow because its measure includes additional microstates that become accessible relative to current values of macroscopic properties."

This may seem counter-intuitive to those who, since Schrödinger's (1945) text, have struggled to become comfortable with the notion that evolution is negentropic. We will illustrate our views using a measure developed by Karreman (1955). To quote Hollinger and Zenzen (1982; see also Van Ness 1983), we will assume that any macroscopic system that is reproducible (aperiodic) enough to be described topologically has an entropy and its entropic behavior can be measured:

> Fortunately there is a well-known way to approximate the statistical entropy for many models, and the ensemble methods of Gibbs extend this approach to any model of thermodynamic interest. The approximation becomes exact in the limit of large numbers of particles provided it is applied correctly. The method involves distributions of particles or other subsystems over their states. Corresponding to the macroscopic data there may be many accessible distributions, and corresponding to each distribution there may be many accessible microstates. If the most probable distribution is overwhelmingly most probable in the sense of claiming the overwhelming majority of accessible microstates, then the entropy of the macrostate can be approximated as the entropy of that most probable distribution which is given by familiar formulas such as the one in equation (12) [see page 54].

There are three aspects of ensemble theory that bear mentioning at this point. First, ensemble measures are used when initial conditions are not known with certainty, but the system being studied is dynamic enough that something of its properties can be estimated by studying the present state of the system. Second, the ensemble is generally characterized in a "phase space" of $6N$ dimensions, where N is the number of elements in the ensemble (usually molecules) and 6 refers to three motion and three spatial dimensions required by quantum theory. An equivalence between thermody-

namic entropy and statistical entropy has been shown at the molecular level (Brillouin 1962). If the system is highly aperiodic, as
biological systems are, the motion variables can be neglected and
the ensemble characterized by the spatial coordinates and the number of elements. Of course, each such approximation and simplification renders the results quantitatively imprecise, but since ensemble
theory is used primarily to gain insight into the qualitative workings
of systems, valid information may be obtained from such an approach. And third, N is generally considered a constant, a static
initial condition. This is a valid assumption only for closed systems,
since open systems exchange matter and energy with their surroundings. Therefore, for characterizing nonequilibrium ensembles, we
need not assume $N = k$ a priori (see Landsberg 1984a, 1984b). In
addition, since nonequilibrium systems exhibit irreversible behavior,
we can assume that N will either remain the same or increase through
time, but will never decrease, even though at any given time, only
a portion of the N number of elements may be functional.

There is a view of microscopic information derived from theories
of dilute classical gases (see Layzer 1977 for a discussion of this
point), for which microscopic information resides in many-particle
correlations. For 100 "particles" distributed over 100 microstates
(and calculated using $k = 1$ and logarithms in base 2 to give units
of bits of information):

$$H_{max} = \log_2 100!$$
$$H = \log_2 100$$

and

$$I = H_{max} - H = \log_2 99! \text{ bits}$$

However, we are not dealing with microscopic information, nor are
we dealing with dilute gases. The nonstatistical, microscopic information above is not an appropriate measure for biological systems.
We wish to follow the macroscopic allocation of components ("particles") into various states. For 100 "particles" behaving in a statistical manner relevant to biology, the maximum possible distribution
at any time is

$$H_{max} = -\sum p_i \log_2 p_i$$

where $p_i = 1/100$; hence

$$H_{max} = -\log_2 1/100$$
$$= 6.644 \text{ bits}$$

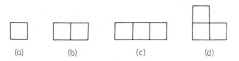

Fig. 2.2.   Four simple geometric figures used to illustrate calculations of macroscopic entropy and information. Vertices represent discrete elements ("particles") composing the structures. Lines connecting vertices represent interrelationships among elements that define structure. For methods of calculation, refer to the text.

If ten "particles" each are distributed in ten different states (such as ensembles of genes distributed over organisms), $p_i = 1/10$ and

$$H = -\log_2 1/10$$
$$= 3.322 \text{ bits}$$

and the macroscopic information is

$$H_{\max} - H = 3.322 \text{ bits}$$

In fig. 2.2a, there are four points in the structure, each equally defining the total ensemble. The resulting probabilities (P_i) are thus all $\frac{1}{4}$ (thus totaling 1), and

$$H_a = -(4)\left(\frac{1}{4}\log_2\frac{1}{4}\right)$$
$$= 2.00 \text{ bits}$$

Now consider fig. 2.2b. There are two classes of points describing the phase space, four signifying a single aspect and two signifying two different aspects simultaneously. The probabilities are 1/8 for the first class and 2/8 for the second class of points. For this ensemble:

$$H_b = -(4)\left(\frac{1}{8}\log_2\frac{1}{8}\right) - (2)\left(\frac{2}{8}\log_2\frac{2}{8}\right)$$
$$= 2.50 \text{ bits}$$

The entropic behavior of the system from state a to state b is

$$\Delta H = 2.50 - 2.00$$
$$= 0.500 \text{ bit}$$

Thus, there has been positive entropy production associated with the increased structural complexity from a to b.

Now consider figs. 2.2c,d. For the former, there are still two classes of points, four indicating a single element and four others indicating two elements. The probabilities are thus $\frac{1}{12}$ and $\frac{2}{12}$, respectively, and

$$H_c = -(4)\left(\frac{1}{12}\log_2\frac{1}{12}\right) - (4)\left(\frac{2}{12}\log_2\frac{2}{12}\right)$$
$$= 2.92 \text{ bits}$$

For fig. 2.2d, there are three classes of probabilities, five indicating single elements, two indicating double elements, and one indicating three elements. For this structure:

$$H_d = -(5)\left(\frac{1}{12}\log_2\frac{1}{12}\right) - (2)\left(\frac{2}{12}\log_2\frac{2}{12}\right) - (1)\left(\frac{3}{12}\log_2\frac{3}{12}\right)$$
$$= 2.86 \text{ bits}$$

The entropy produced by the evolution of ensemble 2.2b to 2.2c is

$$\Delta H = H_c - H_b = 0.42 \text{ bit}$$

and for 2.2b to 2.2d, it is

$$\Delta H = H_d - H_b = 0.36 \text{ bit}$$

Note that in either case, the amount of entropy produced is less than that produced during the evolution of 2.2a to 2.2b. This is because the proportion of new microstates accessioned (i.e., the dynamic entropy) has dropped, that is, proportionately more of the microstates available to the system are historically determined (the historical entropy). Note also that each new structural part in fig. 2.2b–d requires the accession of previously accessed microstates. In fig. 2.2c, two microstates previously accessed once each have been utilized in the incorporation of still more microstates; in fig. 2.2d, two previously accessed microstates are used as well, but one of those had been accessed twice previously, in fig. 2.2a,b. For that reason, what we would call *historical constraint*, there is a smaller increase in entropy from fig. 2.2b to 2.2d, than from b to c. When plotted against time, we see this tendency as a trend toward reducing entropy increase per unit time, or a form of minimizing the rate of entropy production. Thus, the principal of minimum entropy production (Prigogine 1980) appears to have more general properties than strictly thermodynamic ones.

In subsequent chapters we will show that variants of this measure of statistical entropy, or *topological information content* (Karreman 1955), can be used to measure many aspects of the entropic behavior

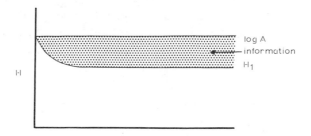

Fig. 2.3. Schematic representation of the relationship between macroscopic entropy and macroscopic information in systems whose macroscopic behavior is due entirely to extrinsic perturbations. Log A defines the maximum entropy of the system, where A is the number of possible states, or complexions, of the system. H is the entropy of the observed state of the system at any time. Macroscopic information is $\log A - H$. (Redrawn from Brooks and Wiley, *Syst. Zool.* 34 [1985].)

of living systems. Because living systems are nonequilibrium systems, the most urgent feature of their behavior will be the evolution of increasingly complex ensembles of microstates behaving in a historically constrained coherent manner.

Note that Karreman equated information content and statistical entropy. Classical communications theory (Shannon and Weaver 1949) also considered entropy equivalent to information. However, measurement theory and more recent views of information derived from communications theory (e.g., Gatlin 1972) considered information to be a measure of the degree of difference between the actual entropy of a system and the maximum possible entropy. Most information theorists have considered information and entropy to be inversely related. For systems approaching equilibrium, increasing entropy is signified by decreasing information. At equilibrium ($P_i =$ 1), there is no information content. One therefore generates information by moving away from equilibrium, and if entropy and information are inverses, this involves the destruction of entropy, or negentropic behavior (fig. 2.3). Thus, biological evolution would be negentropic (see, for example, Mercer 1981). But, we know that entropy production is associated with systems moving away from equilibrium and that negentropic behavior does not occur in reality. Therefore, there must be some forms of entropy production associated with information increases.

Reconsider fig. 2.2. For the ensemble in fig. 2.2a, $N = 4$; for b, $N = 8$; and for c and d, $N = 12$. Maximum statistical entropy is log N, usually designated log A in information theory, where $A =$

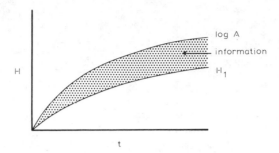

Fig. 2.4.　Schematic representation of the relationship between macroscopic entropy and macroscopic information in systems whose macroscopic behavior is due at least in part to initial conditions constraints. Log A and H are defined as in fig. 2.3; macroscopic information is log $A - H$. (Redrawn from Brooks and Wiley, *Syst. Zool.* 34 [1985].)

the number of "letters" in the alphabet. Maximum entropy for fig. 2.2a is 2.0 bits; for b, it is 3.0 bits; and for c and d it is 3.5 bits. If information is calculated as the difference between log A and H, then

$$I_a = 2.0 - 2.0 = 0 \text{ bit}$$
$$I_b = 3.0 - 2.5 = 0.5 \text{ bit}$$
$$I_c = 3.5 - 2.92 = 0.58 \text{ bit}$$
$$I_d = 3.5 - 2.86 = 0.64 \text{ bit}$$

We find that in open systems, where the irreversible exchange of matter and energy produces increasingly larger ensembles, entropy and information increases can occur concomitantly (fig. 2.4). Note that history is responsible for increasing the size of the ensemble as well as being responsible for the constraints on entropy increases within the system. Note also that a complete description of the system resides in the macroscopic entropy and macroscopic information terms; in accordance with the views of Layzer (1977), microscopic information does not exist or is redundant with the macroscopic entropy.

Information and Entropy

Three critical questions arise at this point and need to be answered by our theory. First, what is the relationship between what we call biological information, or information inherent in living systems, and the mathematical notions of information; and how can that infor-

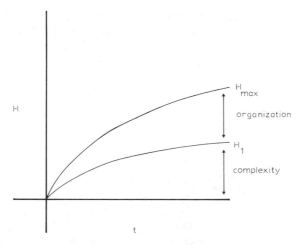

Fig. 2.5. Schematic representation showing that macroscopic behavior in systems having inherent constraints involves both increasing complexity and increasing organization. (Redrawn from Brooks and Wiley, *Syst. Zool.* 34 [1985].)

mation exhibit entropic behavior? Second, what is the source (mechanism or mechanisms) of the time asymmetry in our theory? And third, how can we use Boltzmann's notions of statistical entropy and reject a quantum view of biology?

Information is generally considered only a phenomenological description of the state of a system. In the case of the dynamic shown in fig. 2.5, the information reflects the degree to which initial conditions constraints have kept entropy increases below maximal. It is an assessment of the degree to which the system has failed to become maximally random. That nonrandomness that is transmitted to the next generation as initial conditions for development *constrains* or *informs* the system with regard to possible energy dissipation. In this sense, the information is more than a phenomenological description of the system—it is part of the causal explanation for the system's evolution (Gatlin, 1972, anticipated this relationship). Thus, the phenomenological information is a direct assessment of the inherent information in any biological system having an inherited component. There is a direct link between these two senses of information. Furthermore, since the amount of phenomenological information for any system is dependent on the system's entropy, changes in phenomenological information must be due to entropic changes in the inherent biological information.

Our concept of information is not completely compatible with traditional views of information stemming from communications theory (Shannon and Weaver 1949) or from measurement theory (Brillouin 1962). Collier (in press) has suggested that this is due to shortcomings in the traditional views. The realm of macroscopic information generated by an evolving system constrained by its initial conditions (figs. 2.4 and 2.5) is a form of *free information* as defined by Brillouin (1962). A subset, or special case, of free information is that due to structural organization, and this is called *bound information*. Bound information includes structure held together by forces extrinsic to the system as well as structure held together by inherent forces. Collier termed the bound information that is physically embodied due to inherent constraints *array information*. Array information is physically encoded; thus, it is also a subset of *message information* as construed by communications theory. It is very similar to Gatlin's (1972) concept of *stored information*. When we use the terms *information* or *macroscopic information,* we mean biological array information as defined by Collier.

Information is defined in all these cases as the difference between the maximum possible entropy and the entropy of the observed state of a system. The measure of biological array information, which is a special case of real physical information, is not handled adequately by measurement theory. Measurement theory, in accordance with traditional thermodynamics, relies on external measures of entropy that measure the state of the constraints on the system and that assume the boundary conditions determine the information content of the system. Information is thus seen as an internal property and entropy an external property of a system. Because biological systems become more constrained through time, external measures of entropy would suggest that biological functioning is an entropy-reducing process.

If a system is homogeneous and near equilibrium, external constraints do define the system. But if the system is inhomogeneous and far from equilibrium, as biological systems are, external measures of entropy will be unable to distinguish inherent constraints from external ones and will incompletely describe the state of the system. Our internal measure of entropy shows that biological systems are entropy-increasing with respect to themselves, even if they appear entropy-reducing relative to a static observer measuring entropy externally. Biological array information is not handled adequately by communications theory either, because communications theory has an internal measure of entropy that is unphysical (it can decrease as a result of irreversible processes) and that considers the information content of a system to be external to it (defined in terms of channel capacity). Our internal entropy behaves in a physical manner, and our biological array information is an inherent property of the system.

Biological array information is hierarchical, and the information at any level is generated by probabilistic processes at lower levels, each constrained by a particular set of initial conditions. Within each level, biological array information comprises *stored information,* which is actualized and persistent as organized structure, and *potential information,* which is dissipated by the entropic behavior of the system. In this way, both information and entropy increase in biological systems, in a manner entirely consistent with general models of dissipative structures and of macroscopic behavior resulting from initial conditions constraints.

Coherent biological functioning requires some degree of organization of matter. If, as we claim, this organization is due to initial conditions constraints, and if those initial conditions are provided by the previous generation, there is an inherent time asymmetry in the process. Thus, if proper functioning of biological systems requires accession of information produced in previous time periods, that biological functioning will be a macroscopic phenomenon. The inheritance of biological information, in the sense described in the previous paragraph, ensures that the products of reproduction are organized initially. This will be true even if the inheritance is imperfect, that is, if the flow of information is entropic, and even if boundary conditions affect the system. Thus, under our theory, established mechanisms of inheritance are the primary cause of macroscopic behavior at various functional levels in biology.

Systems constrained by their initial conditions are commonly thought of as completely deterministic and not macroscopic. This being the case, how can we use statistical entropy arguments to represent an initial conditions model? The answer is this: if the number of microstates accessible to a system increases through time, the system's entropy increases. If the phase space containing the microstates accessible to the evolving system also expands, and expands at a faster rate than the increase in the number of microstates occupied by the system, macroscopic information will be produced as a result of the entropic behavior of the system. Initial conditions constraints are responsible for keeping a system from becoming randomly distributed over all possible microstates, producing macroscopic information. Thus, so long as we take into account both the number of microstates over which the system is distributed and the size of the phase space in which the system occurs at any particular point in time, coarse-grained statistical entropy estimates will be appropriate indicators of the macroscopic behavior of a system that is constrained by its initial conditions (see Landsberg 1984a, 1984b). One possible source of confusion about this issue may stem from a form of historical constraint—Boltzmann's use of the word "probability" to refer to macrostates. Brush (1983) stated:

> The fact that a macrostate can be assigned a certain "probability" does not necessarily mean that its existence results from a random process. On the contrary the use of probabilities here is perfectly compatible with the assumption that each microstate of a system is rigorously determined by its previous state and the forces acting on it. We need to use probability measures because we must deal with macrostates corresponding to large numbers of possible microstates. Boltzmann might have avoided the connotations of the word probability by using a neutral term such as "weighting factor."

The remaining chapters will be devoted to showing that the evolutionary dynamics implied by this discussion pertain to all functional levels of biological organization and evolution. We have asserted that biological evolution is a manifestation of a general principle of evolution that also includes thermodynamic evolution. Layzer (1975) and Frautschi (1982) discussed the manner in which the evolution of organized structures from initial chaos in an expanding universe could be reconciled with statistical mechanics. They concluded that the expanding universe comprises a "causal" region in which entropy increases but progressively lags behind the maximum possible increase allowed by the expansion of the "causal" region due to the initial conditions constraints of gravity. The result of this growing disparity between possible and actual is the emergence of organized macroscopic structures. In terms of our model, the "causal region" is $\log A$, and the growing disparity between $\log A$ and H_1 results in biological organization. Layzer predicted that there should be an "arrow of time" ($\log A$), an "arrow of entropy" (H), and an "arrow of history" (I). Information is produced as an entropic phenomenon by any natural macroscopic system whose entropic tendencies are constrained at least partially by its initial conditions, that is, its history. Our view of biological evolution is thus entirely consistent with cosmological evolution.

A similar sentiment has been expressed, in a more reductionistic vein, by Elsasser (1983). He suggested that higher levels of biological complexity, specifically macromolecular structure, could be described by expanding the phase space to include dimensions defined by the higher levels. His general term for this expanded phase space was *polymeric space;* it was to be described by "boxes" analogous to the quantum dimensions and "balls" analogous to the number of molecules. He asserted: "If a set of abstract entities called "boxes" and "balls" has been clearly defined the validity of the corresponding Second Law, providing internal changes of the system do occur, is inescapable. The main problem lies therefore in the adequate definition of boxes and balls, which may be done in various ways under the complex conditions of biology."

Elsasser observed that if a system occupied only part of its phase space, any internal change in the system would be entropic. All that is needed is a mechanism for generating partially filled phase space. Elsasser suggested that natural selection might act as such a mechanism. However, our view demonstrates another means of achieving the same ends. The value log A may be seen as a generalized assessment of the size of the "phase space" for any system: H is a measure of the space occupied by the system. So long as there is a difference between log A and H any internal change in the system must be entropic.

Landsberg (1984a, 1984b) examined macroscopic behavior in terms of three properties, *super-additivity* (S), *homogeneity* (H), and *concavity* (C). Super-additivity is defined as

$$S_{(a + b)} \geqslant S_{(a)} + S_{(b)}$$

where $S_{(a+b)}$ is the entropy of a merged system composed of two subsystems, a and b, with partial entropies, $S_{(a)}$ and $S_{(b)}$. The merging of two homogeneous subsystems should not change the total entropy, and the merging of two inhomogeneous ($\overline{H}$) subsystems should increase the overall entropy. Thus, Landsberg concluded, no naturally occurring macroscopic system can violate super-additivity ($\overline{S}$). Violation of this property would occur if two subsystems were merged and the resulting entropy were less than the original, which is an unphysical result.

Homogeneity is defined as the state in which the entropy of a system as a whole is unaffected by partitioning; hence

$$S(aX) = aS(X)$$

Whenever homogeneity is violated ($\overline{H}$), the entropy associated with the system becomes *nonextensive,* meaning, among other things, that there is no requirement that macroscopic behavior in such systems achieve a maximum entropy equilibrium. This allows the possibility of simultaneous increases in entropy and order. As indicated above, neither homogeneous nor inhomogeneous systems violate super-additivity.

The third property is concavity. Landsberg and Tranah (1980) suggested that the second law is usually taken to imply that entropy is a concave function of any system X, such that for a constant λ, $0 \leq \lambda \leq 1$

$$S[\lambda X_1 + (1 - \lambda)X_2] \geq \lambda S(X_1) + (1 - \lambda)S(X_2)$$

This relationship (C) will hold under conditions of super-additivity when the system is homogeneous and will be violated $(\overline{C})$ when the system is inhomogeneous.

Given these observations, Landsberg (1984a, 1984b) concluded that naturally occurring macroscopic systems conform to one of two sets of properties; they are either (S, H, C) or they are $(S, \overline{H}, \overline{C})$. The other six possible combinations of properties imply unphysical results. The first class of systems is represented by ideal gases, in which macroscopic behavior is determined solely by the local forces, or boundary conditions. It would appear that the viewpoint espoused by Prigogine and Stengers (1984) not only assumes that close-to-equilibrium behavior holds for far-from-equilibrium phenomena, but also assumes that all such systems are made up of homogeneous parts. Landsberg characterized the second class of systems as those controlled by long-range forces, or initial conditions. Inhomogeneous mixtures will not hold together as systems in the absence of long-range forces. Suh systems can exhibit simultaneous growth of entropy and information (I, where $I = H_{max} - H$) or order (Q, where $Q = 1 - H/H_{max}$), so long as the number of accessible microstates grows faster than the system's own entropic growth can distribute the system among the microstates. This leads to the production of higher levels of organization with inhomogeneous partial entropies, which can best be described by using coarse-grained estimates of entropy.

The coarse-grained entropies that we will present in the following chapters can be related in principle to strictly thermodynamic considerations in the following manner (modified from Landsberg 1984b): the statistical entropy of a given system of n accessible microstates at a given energy, if $p_i = 1/n(t)$ (implying homogeneity), is

$$S = k \ln n(t)$$

For systems that violate homogeneity, $p_i \neq 1/n(t)$, thus

$$S = -k \sum p_i \ln p_i$$

Under such conditions of inhomogeneity, we would expect the accessible microstates to range over a set of varying energies. If n is the number of accessible microstates over a range of energies (E_i), then $p_i = \exp(E_i/kT)/Z$ and

$$S = k \ln Z + U/T$$

where $Z = \Sigma \exp(E_i/kT)$ and $U = \Sigma p_i E_i$.

Our biological theory converges strongly with the above recent developments in physical theory dealing with macroscopic behavior of systems controlled by their initial conditions. We postulate that heredity and reproductive ties exert an analogous kind of influence on biological systems that gravity exerts on astronomical systems. As anticipated by Layzer, Frautschi, Elsasser, and Landsberg, we believe that biological systems and self-gravitating astronomical systems are members of the same class of macroscopic systems. Although developed initially in analogy with thermodynamics, we believe our theory is more than just an analogy.

Hereafter, we will often use the term *complexity* to refer to the statistical entropy of the system and the term *organization* to refer to the information content (*sensu* Gatlin 1972)(fig. 2.5). This is necessary to avoid confusion when we discuss the behavior of biological information, which can show entropic behavior and generate more "information." Thus, the complexity of a given system is signified by the height of the H curve at a given point, and the organization is signified by the difference between the log A and the H curves at the same point. The general dynamic established by our discussion and supported by more general concepts of evolution is one in which both complexity and organization increase through time (i.e., figs. 2.4 and 2.5). The amount of increase in complexity (entropy) declines as the historical burden of constraint grows. Thus, we can see a direct connection between historical constraints and minimum entropy increase. The increase in organization occurs as actual complexity increases lag farther and farther behind the maximum possible increases; thus, it is also a time-dependent or history-dependent phenomenon. The degree to which the inherent dynamics of complexity increase result in increasing organization is the degree to which the evolving system is self-organizing. Our proposed evolutionary dynamic thus links, through inherent historical constraints, both the principle of minimum entropy production (Prigogine 1980) and concepts of self-organization or *synergetics* (Haken 1978). This gives us hope, along with the sentiments of Landsberg (1984a, 1984b), Elsasser (1983), Frautschi (1982), and Layzer (1975), that there really is no fundamental difference between strictly thermodynamic and statistical mechanical views of entropy and of biological evolution.

Shimizu and Haken (1983) recently suggested that the strictly thermodynamic views of Prigogine and his coworkers accounted for emergent complexity and organization strictly on the basis of perturbations by extrinsic forces (energy fluxes). That is, organization is imposed by the boundary conditions. According to them, the synergetic view explained emergent complexity and organization as manifestations of the self-organizing capacities of living systems, or organization emerging from the initial conditions. While the model we have presented suggests that evolution is ultimately self-organizing, it does not preclude the actions of proximal, "perturbing" influences. We believe a complete theory of evolution will be one which accounts for the relative influences of each mode of causality.

We are not in a position to demonstrate a rigorous identity between the kind of self-organization emerging from our view of biological evolution and the self-organization of biological macromolecular structure as discussed by Eigen (1971). We would note again the similarities between our general dynamic and the views put forward by Elsasser (1983), who was clearly thinking in macromolecular terms. Furthermore, we would draw attention to the cell biological work of Kucias (1984 and references therein) for an empirical link between our views and Eigen's. If our theory is headed in the right direction, such a link between molecular self-organization and self-organization at higher functional levels should be forthcoming.

Summary

We may now provide some answers to the questions posed in this chapter and present the framework for our theory of biological evolution.

What is the relationship between biological evolution and thermodynamic evolution? Both are particular manifestations of a general phenomenon of entropy production in physical systems exhibiting time-dependent changes. The second law is more than a law of thermodynamics—it is the natural law of history.

Why is there order and not chaos in the living world? Because living systems, organisms and species, are individualized dissipative structures (1) exhibiting finite information and cohesion, (2) maintaining themselves through irreversible dissipation of matter and energy, and (3) existing in an open energy system. Why do we find a particular kind of order, that is, individual functional units and hierarchically related species? Because (1) entropic behavior holds for all classes of irreversible processes, even those not strictly thermodynamic, (2) replication is imperfect and cohesion can be broken, and (3) there are initial conditions (historical and developmental) and boundary conditions (natural selection) constraints on the entropic behavior of information and cohesion. And why do particular organisms look the way they do? Because each is the product of a unique history in which intrinsic constraints restricted the number of kinds of variants produced and extrinsic constraints eliminated those realized variants that were not adequate for survival. Biological evolution is not a teleological process, nor is it a process that requires us to postulate that better adapted variants occur randomly and are "selected" because of their functional efficiency in a given environment. Rather, the most urgent property of living systems as entropy systems is historically constrained structural evolution regardless of the environment. Evolution is survival of the adequate, not of the most fit.

We suggest an alternative theoretical framework for biological evolution, based on four principles:

1. *The principle of irreversibility.* By this term we mean to emphasize the connection between biological evolution and the thermodynamics of irreversible processes as entropic behavior. The components of living systems relevant to biological evolution are therefore those which exhibit irreversible (i.e., entropy-producing) behavior. There are four such components: (1) metabolism and homeostasis; (2) growth and differentiation, or ontogeny; (3) reproduction and population differentiation; and (4) speciation, or lineage splitting.

If biological evolution is a nonequilibrium entropic process operating through the four processes listed, we should be able to discern entropy production in measuring them. These measures would be either physiological, based on metabolic rates, or structural, based on measures of statistical entropy and information, or both.

Nonequilibrium systems exhibit two features of primary interest to evolutionary theory. The first is that they embody their own history in their causal makeup and mechanical behavior. Therefore, documenting the hierarchy of evolutionary relationships is of paramount importance in providing evolutionary explanations. But because history is so intimately entwined in the dynamics of the system, discovery of the hierarchy should be feasible based on observed attributes of contemporary systems (see chapter 5). The second is that nonequilibrium systems tend to evolve toward states of minimum entropy increase, or minimum entropy production. These states are historically emergent and can be discerned retrospectively, but cannot be predicted from initial conditions.

2. *The principle of individuality.* Entities that evolve must exhibit spatiotemporal continuity and some intrinsic boundary conditions. In short, they must be individualized (Ghiselin 1974; Hull 1976). We noted earlier that Goldschmidt (1952), Hennig (1966), and Riedl (1978) recognized that any irreversible process operating on discrete, or individualized, entities produces a hierarchy. If we have a hierarchy, there must be individualized components of the process responsible for its existence. This means that genealogical origins are an essential part of any species along with a unique historical burden of inherited information. This will be our clue to a method for recovering the hierarchy. Alternatively, classes of phenomena, like frequencies of particular alleles in populations and functional attributes of organisms, lack spatiotemporal continuity and thus cannot be causal agents, although they may be outcomes, of evolution.

Individuality is also an important concept to consider when one tries to relate biological evolution to general causal laws. Popper (1965) was among the first to recognize that a difficulty existed because the evolutionary hierarchy is a singular statement and not a universal statement. However, this does not mean that there are no general causal laws governing evolution. If evolutionary units are individualized physical systems, such as dissipative structures, their historical fate may be understood in terms of causal mechanisms governing the behavior of dissipative structures. The principle of individuality thus provides a bridge principle (*sensu* Hempel 1966) between the singular statement of phylogeny and the irreversible, time-dependent dynamical processes of dissipative structures in general.

The principle of individuality is important for a third reason. It provides an ontological connection between organismal-level and population- and species-level processes. As we suggested in chapter 1, the success of neo-Darwinism has been due, in great measure, to the shift from individual (also called "typological" or "essentialistic") thinking to population thinking in terms of evolution. This reinforced the idea that different functional levels were ontologically distinct, and thus only one could be fundamentally relevant to evolution. However, if the different functional levels of biology can be shown to be connected ontologically, then a general theory of mechanism should be sought and should be possible to discover.

3. *The principle of intrinsic constraints*. Intrinsic constraints are of two kinds—historical and developmental. The recent attempts to improve evolutionary theory (Gould 1980; Dover 1982; Wicken 1980; Ho and Saunders 1979; Saunders and Ho 1980) have all stressed the need to consider and to discover internally generated factors governing the production of evolutionary novelties upon which natural selection operates. We have seen in chapter 1 that this is one of the great recurring themes in evolutionary theory. The converse of such mechanisms for generating variation is just as important, however; what *constraining* effects do historically and developmentally determined states of living systems have on the production of novelties, independent of the effects of selection? If evolution is a manifestation of a general law of entropic behavior, we do not have to provide any special explanation for increasing variation. Rather, we must explain why not all conceivable variants are realized. This principle asserts that initial conditions are causally important in evolution, even though they are not sufficient to explain particular end states.

4. *The principle of compensatory changes*. The principle of compensatory changes can be derived in the following manner. As noted before, for open systems

$$dS = d_eS + d_iS, \; d_iS > 0$$

If the system is close to (thermal) equilibrium, initial conditions can be neglected, d_iS comprises only energy dissipated from the system into the environment, and dS can increase or decrease depending on the relative absolute magnitudes of d_eS and d_iS.

Systems far from equilibrium are highly constrained by their initial conditions, and for them d_iS comprises multiple partial entropies. Those partial entropies associated with boundary conditions effects (such as metabolic rate, population size, and number of different species) can fluctuate up or down, depending on the amount of available energy used, so long as there is an abundance of free energy. Those partial entropies associated with initial conditions effects (information and cohesion) can only stay the same or increase since each state is dependent on a historically actualized predecessor state, unless time runs in reverse. Under our theory, fluctuations in partial entropies associated with boundary conditions effects will affect the rate at which the partial entropies associated with initial conditions effects increase. Hence, any change in initial conditions that decreases the apparent entropy of the system will be compensated by an increase of equal or greater magnitude in other portions, or the system will break down and become nonfunctional (i.e., it will go to thermal equilibrium).

Since the beginning of evolutionary thinking, biologists have labored under the belief that they have to explain why there are so many species. But that is not the case at all. Given the detectable genetic complexity and possible combinations, we have to explain why there are not more species than there are. Evolution is a process that slows down the entropy decay of lineages, minimizing their entropy increases. This suggests that, as the interplay of information and cohesion, biological evolution should exhibit an intrinsic tendency toward efficiency or parsimony, which in turn should relate to the principle of minimum entropy production.

Any evolutionary novelty must integrate with the total developmental program of any organism in which it occurs. If the novelty affects other parts of the organism during development, those other parts must be capable of compensating for the presence of the novelty, or the organism will not develop properly. Additionally, new species are initially subsets of ancestral species and, as such, are less variable than the ancestor in at least some characters. Speciation, the historical splitting of lineages, thus tends to "reduce" ancestral variation by partitioning it into different lineages. And yet, if new variation did not occur in descendent species, evolution would have ceased long ago after all variants of the first species were isolated into discrete lineages. This has not happened; new variants show up in descendent species. The evolutionary process, as indicated by the historical hierarchy, appears to be one that involves the generation of determinate factors (traits characteristic of all members of a particular species, the historical entropy) and of stochastic factors (new variation in descendent species, the dynamic entropy). The discovery of an intrinsic (= genetic) basis for compensatory changes would provide evidence that there is always a pool of new microstates accessible to any given species, providing biological evolution with its own directionality and continuity. In chapter 3 we will discuss evidence that such a tendency does exist.

In the following chapters, we will attempt to show how these principles are expressed in different functional levels of biology. We will suggest research programs designed to test our theory and provide evidence of current findings to support it.

· 3 ·

Ontogeny, Morphology, and Evolution

Differences we observe among species have their origins in differences in ontogenetic programs that have evolved in the past. Thus, developmental biology should be a crucial field in evolutionary biology; indeed, it was crucial in the nineteenth century. But as Hamburger (1980) and Raff and Kaufman (1983) have noted, developmental biology has played a minor role in twentieth-century formulations of evolutionary theory. Raff and Kaufman suggested that the lack of a synthesis between developmental biology and evolution was due to (1) rejection of the biogenetic law by experimental embryologists and (2) a schism between the fields of genetics and embryology. Attempts at a synthesis, notably by Goldschmidt (1940), failed because of a lack of integration of genetics and ontogeny. This does not mean, of course, that there was no appreciation for the relationship of development and evolution. Such workers as Garstang (1922), Huxley (1944), Waddington (1957), and de Beer (1958), dealt with the relationship of ontogeny and evolution. Renewed interest, evidenced by both developmental biologists (Kauffman 1983; Raff and Kaufman 1983) and evolutionary biologists (Løvtrup 1974; Riedl 1978; Alberch 1980, 1982, 1985; Alberch and Alberch 1981; Gould 1980) has deepened our understanding of the relationship. Yet a unified integration has not yet been reached.

Ontogeny is an irreversible process and thus is a candidate for integration into a hierarchy of such processes. We suggest that ontogeny, growth, and differentiation provide the interface between strictly thermodynamic entropy production and structural (i.e., statistical) entropy production produced by phylogenetic descent, because it is the process in which energy flows and information changes interface directly. Thus, it is of particular interest to our theory because developmental biologists can measure entropy production directly, either in the form of energy flows (metabolic rates) or in terms of genetic and/or structural changes.

To integrate with the entropic view of evolution, ontogenesis must satisfy several criteria. (1) It should be a nonequilibrium phenomenon in which each ontogenetic unit (organism) is a dissipative structure fluctuating from one steady state to another through time; (2) organisms should exhibit a tendency toward minimum entropy production, expressed either as a rate of energy dissipation or degree of structural change as development proceeds; (3) the dynamics of ontogeny should be at least partially self-organizing, in which each new steady state is causally related to the preceding one(s); and (4) ontogeny should be a process characterized by alternate dissipative pathways and compensatory changes. We should find evidence of symmetry breaking, aperiodicity, increasing complexity and reproducibility, and bifurcations in the mechanics of ontogeny.

At issue are two general questions. First, does ontogeny exhibit lawlike behavior in individual cases and in evolutionary transformations? Second, what are the relative roles of factors intrinsic to the organism and extrinsic to the organism in development and the evolution of different developmental programs? The two questions are related, because every organism develops within an intimate relationship with its environment. Traditionally, this has been interpreted as an indication that the environment (i.e., extrinsic factors) is the determinate cause of any orderly behavior of biological processes, including ontogeny (see chapter 1). However, there is no way any organism can avoid intimate interactions with the environment. Thus, such observations cannot be taken as de facto evidence that the environment is determinate. Rather, they are indications that the environment is a necessary part of the boundary conditions for development. In addition, some developmental stages will be expressed even in different environments, ensuring faithful replication of heritable traits. This raises the possibility that there are deterministic factors operating in the expression and evolution of ontogenetic programs that are beyond environmental control.

We recognize three paired classes into which traits produced by ontogenesis may be placed for the purposes of discussing evolutionary mechanisms (table 3.1). Every trait satisfies one of the two possibilities for each class. For example, the seasonal switch from thick winter coats to thin summer coats in many mammals is heritable, reversible, and environmentally dependent. The switch from

Table 3.1 Classification of Traits Produced by Ontogenesis

	Traits	
Class	Possibility 1	Possibility 2
I	Heritable	Nonheritable
II	Irreversible	Reversible
III	Environmentally independent	Environmentally dependent

white winter coats to brown summer coats in ermine in the arctic is heritable, cyclically reversible, and environmentally independent. Skin cancer in humans is nonheritable, irreversible for the cell lineage involved, and environmentally dependent.

Multicellular organisms are the result of a sequence of developmental events. Organisms look similar (barring convergence) to the extent that the sequence of developmental events is the same. For any multicellular organism, the sequence of events is partially determined before conception by maternal RNA already present in the egg cytoplasm. Development proceeds after fertilization as a sequence of cell divisions and cell differentiations. The entire process gives the impression of order, and thus ontogeny seems to be a very deterministic process. We suggest that ontogeny is orderly because it is a nonequilibrium phenomenon, and such phenomena show orderly increases in structural complexity as one manifestation of entropy production. It is not only orderly, but also predictable. Indeed, we can even predict what will result from changing the process, as in the bithorax mutant of *Drosophila*. We suggest that it is predictable because the information changes involved are also part of a nonequilibrium phenomenon that interfaces with energy flows during ontogeny.

Ontogenesis as a Nonequilibrium Phenomenon

Perhaps the first critical attempt to examine development, growth, and aging of organisms with respect to the principles of nonequilibrium thermodynamics was presented by Prigogine and Wiame (1946). According to their work, the thermodynamics of linear irreversible processes is applicable to ontogenesis. They assumed that ontogenesis is a process of continuous approach of the organism to a final stationary state (adulthood), accompanied by a decrease of the spe-

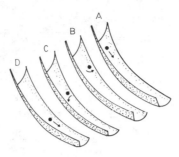

Fig. 3.1.   Schematic model of the deviation of a developing organism from the steady state and its approach to the final stationary state. This is illustrated by a sphere rolling down an inclined trough. At any point in time, the sphere may deviate from the steady state (*B*) under the effect of an external force and then return to the bottom of the trough (*C*) and continue its motion to the final stationary state (*D*). The movement of the sphere along the bottom (*A* and *D*) corresponds to the constitutive approach of the organism to the final stationary state; deviations of the sphere from the current steady state and its return (*B* and *C*) correspond to inducible behavior. (Modified from Zotin 1972.)

cific rate of entropy production of the specific dissipation function ψ. The stationary state is characterized by a minimal and constant rate of entropy production. Zotin (1978) equated ψ with the intensity of heat production (g) or of respiration (g_{0_2}), such that

$$\psi = g = g_{0_2}$$

This enables one to assert that the Prigogine-Wiame hypothesis predicts that there will be a continuous decrease in the intensity of heat production or respiration during ontogenesis. Zotin (1972) asserted that available data for rates of heat production and respiration in various animals supported the Prigogine-Wiame hypothesis.

Zotin and Zotina (1978) also suggested a number of improvements for the Prigogine-Wiame hypothesis. For example, they noted that most animals at rest exist in a steady state with minimal intensity of heat production—the basal metabolic rate. Under the influence of external or internal factors (such as very cold weather or exercise, respectively), organisms may depart from that steady state; however, when the perturbing factors cease, the organism returns to its normal metabolic rate. Such reversible deviations from the current steady state were termed *inducible* processes. They contrasted inducible processes with *constitutive* processes, those which are irreversible approaches to the final steady state. A schematic representation of the differences between constitutive and inducible processes is shown in fig. 3.1.

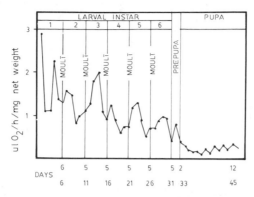

Fig. 3.2.   Respiration intensity of larvae and pupae of the American white butterfly at different stages of development. (Modified from Zotin 1972.)

Inducible processes may be short term, with deviations from and returns to the steady state occurring within the temporal duration of a given steady state (*inducible pulsed processes* of Zotin 1972). They may also be relatively long-term deviations that persist during a phase change from one steady state to another before returning to a different steady state from the one originally perturbed (*inducible adaptive processes* of Zotin 1972).

Zotin and Zotina (1978) suggested that thermodynamic characterization of ontogenesis required a nonlinear approach, beginning initially with

$$\psi = \psi_{(o)} + \psi_{(i)} + \psi_{(a)} + \psi_{(k)}$$

where $\psi_{(o)}$ is the basal metabolism, $\psi_{(i)}$ the inducible pulsed processes, $\psi_{(a)}$ the inducible adaptive processes, and $\psi_{(k)}$ the constitutive processes. Because the different components of ψ proceed at different rates and persist for different periods of time, Zotin and Zotina suggested that ontogenesis involved multiple steady states and a series of fluctuations from one to another in an irreversible sequence; thus, we would expect a nonlinear, or fluctuating, drop in rate of entropy production during ontogenesis (e.g., figs. 3.2 and 3.3). Zotin and Zotina (1978) summarized their findings thusly: "It follows from the Prigogine-Wiame theory, as well as from the experimental results presented above, that the processes of development, growth and aging are all accompanied by a decrease in the specific dissipation function of the system, interrupted only by inducible pulses and adaptive processes."

They asserted that every organism must begin development far from its final steady state; otherwise, we would not see a steady

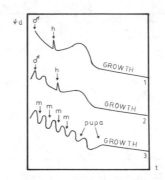

Fig. 3.3.   Scheme of inducible and constitutive processes during development and growth in various groups of animals. *1* = Birds and mammals; *2* = fish and amphibia; *3* = insects; *m* = molt, *h* = hatching or birth. (Modified from Zotin 1972.)

decrease in rate of entropy production. Further, they concluded that at some point prior to the initiation of development there must be a period of increase in rate of entropy production driving the potential organism to the state of a high rate of entropy production. Zotin and Zotina suggested that this occurred during oogenesis, when an increased rate of entropy production in germ cells could be accomplished by coupling the process to a decrease in rate of entropy production in the rest of the organism. Any inducible increases in rate of entropy production following initiation of development were due to energy fluxes from sources external to the organism.

Lurie and Wagensberg (1979) expressed a modified view, suggesting that the nonequilibrium nature of ontogeny was due to the effects of "external constraints" that "are maintained constant from the moment of conception onward":

> In this view, fertilization would transmit the external constraints to the new system and establish its initial configuration as a perturbation or "fluctuation" with respect to the adult stationary state. . . . We take the view that the deterministic aspect of evolution—which is operative at the level of the organism as a whole—may be described by linear thermodynamic laws. This deterministic evolution represents adaptation to externally imposed constraints.

We are in fundamental agreement with the sense of these treatments, but our perspective differs slightly with regard to the nature of (1) the "external" constraints, (2) that to which the organism is adapting during ontogeny, and (3) the way in which an organism achieves its far from equilibrium state.

All dynamic systems, including strictly thermodynamic ones, derive their coherence from interactions, both among the components of the system and between the system and its environment. The word *environment* can mean different things, however. For example, the "environment" of a zygote's nucleus is the cytoplasm of the cell, which is maternal material. The "environment" of the zygote may be, at least initially, parental material or derivatives of parental material (e.g., endosperm in flowering plants) or the physical environment (external fertilization). In multicellular organisms, the "environment" of any one cell during ontogeny becomes progressively less like the parental environment and more like the rest of the cells comprising the organism itself. In one sense, this is the essence of emerging individuality. Overlaid on such considerations is the realization that the nonorganismic "environment" may influence development as well. In short, we recognize that ontogeny is *effected* by interactions with an organismic environment and may be *affected* by a nonorganismic environment as well. We assert that a developing organism adapts first to its history (parental-derived material) and then to itself. It responds primarily to "external" stimuli that are nonetheless organismic in nature; hence, we refer to them as *intrinsic constraints* even though the stimuli are in some sense "external." Those stimuli, or constraints, change throughout development—they are not constant from the time of conception onward.

The process of adaptation, as we construe it in this context, requires boundary conditions stimuli ("perturbations") and initial conditions constrained responses. For development, the initial conditions are those established by fertilization. However, some of the conditions established at that point act as boundary conditions by forming the immediate environment of the other initial conditions. We consider macroscopic behavior constrained by initial conditions to result in *self-organization* and macroscopic behavior constrained by boundary conditions to result in *imposed organization*. Ontogeny embodies an additional macroscopic effect; some of the boundary conditions are self-generated and changing, leading to what we will term *self-imposed organization*. Any developmental events resulting in entropic change that lags behind the growth of the boundary conditions will produce emergent self-organization. Those events characterized by entropic changes that are more rapid than the growth of the boundary conditions will be slowed by the boundary conditions as they approach those limits. This produces self-imposed organization by means of feedback mechanisms. We view ontogeny as a process of emergent and self-imposed constrained entropic increases. The emergent self-organizing capabilities are associated with modern concepts of preformation, whereas self-imposed behavior is associated with concepts of epigenesis.

The distinction between our views and those of Prigogine and Wiame (1946), Zotin and Zotina (1978), Lurie and Wagensberg (1979), and Ho and Saunders (1979) are important when we consider the manner in which an organism achieves its nonequilibrium status. We agree that gametes and zygotes are initially in states close to the equilibrium defined by the parental state because gametes are part of the ensemble of cells defining the parent. However, the parental "equilibrium" is a far from equilibrium steady state and not thermodynamic equilibrium (achieved only at death). Therefore, there is no requirement for a zygote to be "perturbed" from near thermodynamic equilibrium. Using Hollinger and Zenzen's (1982) terminology (see also chapter 2), we view the parental steady state as an old equilibrium plateau to be left behind by the developing organism rather than a static equilibrium from which the organism must be forced initially, only to return.

Following the general dynamic suggested in chapter 2, we view the zygote as initially occupying a state near that of the parent but quite removed from the equilibrium point defined by its own potentials. Fertilization thus removes gametes from parental ensembles and defines a new equilibrium point, toward which the developing embryo spontaneously moves, in accordance with the second law. It thus leaves the parental equilibrium behind. The rate at which this occurs is quite rapid early in development when stimuli provided by the parents provide the greatest control over development (i.e., it is adapting to its history) and when the information system of that new "macrostate" begins accessing microstates at a high rate. This initial rate of entropy production is high relative to the parental equilibrium, but not necessarily higher than the parental rate of entropy production early in its own ontogeny. What is being left behind is the parental equilibrium, not the parental dynamics defined by history and passed on through inheritance. As the organism grows more complex, it begins adapting to the environmental context it provides itself as well. The actual microstates accessed in development are constrained more and more by those accessed previously as well as by the finite response capabilities of the genetic system of the organism. The actual increase in complexity and the rate of thermodynamic entropy production thus begin to lag behind the maximum possible. When the organism no longer responds to "environmental" stimuli by accessing new microstates, its rate of thermodynamic entropy production (*sensu* Prigogine) will be minimal and its complexity will have reached a new equilibrium plateau (both defined as asymptotic approaches to a stable state). The organism has thus exhibited spontaneous entropic decay from its parental "equilibrium," but its actual complexity (H from chapter 2) is much lower than its maximum possible (log A). The organism is thus *organized* (log $A - H > 0$). Since that organization results from historical and developmental constraints provided by the organismic environment, the organism is *self-organized*.

As an analogy, consider the discussion of diffusion by Hollinger and Zenzen (1982; see also chapter 2). Diffusion of heat through water to an equilibrium can be slowed by molecular collisions to such an extent that organized structures (convection currents) form in the water. The dissipation of information can be slowed by historical and developmental constraints to such an extent that a complex and self-organized organism results.

Given this view, let us now return to the equation of Zotin and Zotina (1978):

$$\psi = \psi_{(o)} + \psi_{(a)} + \psi_{(i)} + \psi_{(k)}$$

The terms $\psi_{(i)}$ and $\psi_{(a)}$ refer to temporary (i.e., reversible or periodic) deviations from a steady state, either during or between fluctuations from one steady state to another. Following Hollinger and Zenzen (1982; see also chapter 2), we conclude that such processes do not result in the accession of new microstates and therefore are not entropy-producing processes. While we admit that it is possible to observe and measure such phenomena as manifested by living organisms, they are not part of the causal explanation of ontogeny in general. The equation thus reduces to

$$\psi = \psi_{(o)} + \psi_{(k)}$$

Zotin noted that one could equally well explain the entropy change in a system by noting the origin of entropy in the system or the fate of entropy flowing through the system. He equated Prigogine's work with the first view and his own with the second. Thus, it appears that to Prigogine

$$d_iS = \psi_{(o)}$$

whereas to Zotin

$$d_i S = \psi_{(o)} + \psi_{(k)}$$

We conclude that ontogenesis is a nonequilibrium phenomenon and that a physiological basis for measuring the thermodynamic behavior of developing organisms can be developed by equating changes in metabolic rate with $d_i S$ of Prigogine or $\psi_{(o)}$ of Zotin.

Our theory explains the change from a high to a low metabolic rate during ontogeny, as does the theory of Prigogine and Wiame. How might we decide which is the more appropriate view without assuming one or the other class of macroscopic models a priori? One possible means is use of Landsberg's (1984a) formulation of super-additivity. This property asserts that the entropy of a system formed by merging two subsystems is greater than or equal to the sum of the entropies of the subsystems before merging:

$$S_{(a + b)} \geqslant S_{(a)} + S_{(b)}$$

If the system is controlled only by its boundary conditions, then the property of homogeneity must also be satisfied and

$$S_{(a + b)} = S_{(a)} + S_{(b)}$$

Therefore, if a zygote begins development at a high rate of entropy production, under a boundary conditions model the gametes must be homogeneous subsystems, each with a high metabolic rate equal to half the metabolic rate of the zygote.

If the system is controlled by its initial conditions, then

$$S_{(a + b)} > S_{(a)} + S_{(b)}$$

and homogeneity is violated. We would predict that gametes are not homogeneous subsystems and that the newly fertilized zygote will always exhibit a metabolic rate greater than the sum of the metabolic rates of the gametes. That this is the case is virtually a biological truism. Therefore, if the high metabolic rate exhibited by developing organisms early in ontogeny is not established until fertilization, we can say that ontogeny behaves as a macroscopic process controlled by its initial conditions.

If metabolic changes and structural changes are both reliable indicators of the entropic behavior of developing organisms, we should expect to find super-additive changes in gene activity (the link between metabolism and structure, or between thermal and statistical entropy) at fertilization. Roughly, the number of working genes producing heterogeneous products and raising the entropy in a zygote

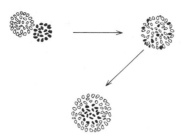

Fig. 3.4.   Equilibrium self-assembly (immiscible liquids, or sorting out and engulf-ment in assemblies of cells). (Modified from Harrison 1982.)

should be greater than the sum of working genes in the gametes. In sea urchins, gametes have approximately 5000 working genes, whereas zygotes have approximately 50,000 working genes (Galau, et al. 1976), substantially more than the sum of the working genes in the gametes. Furthermore, during the course of ontogeny the rate of gene activity (number of working genes per cell) falls in a manner paralleling the changes in metabolic rate. However, as the organism develops and grows, the total number of working genes in its body increases. To have a sense of this for acellular organisms, consider a genealogically continuous population of protists as the "body" of an individual organism. The decreasing rate of activity in these pro-cesses indicates decreasing rates of entropy increase and not de-creasing entropy.

Harrison (1981) defined morphogenesis as ". . . the creation of a complicated shape out of a simpler one by chemical processes in living organisms." Harrison (1981, 1982) also asserted that a major question in the study of causal factors in morphogenesis is deter-mination of how pattern formation should be analyzed. Should it be in terms of geometric structure, that is, the fitting together of pieces in a building-block manner to produce a large-scale structure? Or, should it be considered in terms of equilibrium self-assembly? For the second case, morphogenesis would be best explained as a pro-cess in which the organism's shape emerges as it achieves minimum free energy (thermodynamic equilibrium) through the disposition of intercellular forces. In this type of model, differential adhesion among cells would be due to (or at least analogous to) the equilibrium configurations attained by two immiscible liquids sorting each other out and one engulfing the other (fig. 3.4).

Harrison rejected both the structuralist and equilibrium approaches as inappropriate because they are based on a static rather than a dynamic view of nature. Or, as Harrison (1982) stated, referring to the equilibrium view:

> This picture of cell-as-molecule is a very good start, which needs a lot more attention from physical chemists to establish its scope and limitations. For example, a cell is never, in respect of what is going on inside it, at equilibrium so long as it is alive. Sorting-out or engulfment occurs much too rapidly to be attributable to Brownian motion of such large things as cells. Nevertheless, the active movement of cells may in some instances be random, and the use of equilibrium concepts for their behavior in assemblies will then be valid, with appropriate careful redefinitions of the equivalents to some thermodynamic properties of molecular assemblies.

Harrison (1982) then chose as most appropriate a third possibility, that of chemical kinetics:

> It is possible for a structure, essentially heterogeneous, to arise and be maintained by chemical reactions and transport processes, when the equilibrium state would be homogeneous. Kinetics alone then determine the gross morphology of the system. I would like to call such things "kinetically maintained structures." Some thermodynamicists . . . refer to them as "dissipative structures."

Despite terminological differences, Harrison clearly identified nonequilibrium thermodynamic processes as the basis for causal explanations of morphogenesis. He also stated the growing perception among developmentalists that "the essence of any morphogenetic theory, then, must be a mechanism for symmetry breaking." What is meant by symmetry breaking? In its simplest sense, consider a system in which two different substances are being produced in equal amounts. Symmetry occurs at equilibrium, when the concentrations of each substance are equal, $[A] = [B]$. Symmetry breaking occurs when the system ("$A + B$") is pushed away from the equilibrium point by an increase in rate of production of either A or B relative to the other. For example, the ratio of A to B is $2/1$ at a given time, and A is being produced four times faster than B. The system is driving toward total A. Such a phenomenon is called *amplification*. In such a system, it is possible to use $(A - B) = X$ as a measure of how the system is developing, where X = the asymmetry itself and the system moves away from equilibrium in the following manner:

$$\frac{dX}{dt} = (k_f S/P)X$$

where $P = (A + B)$, $X = (A - B)$, S = the substrate, and k_f = a flow rate. In more complicated models of morphogenesis, P is termed the *morphogen* and is a measure of the displacement from equilibrium between at least two substances. To recognize a "morphogen," then, one needs two or more substances, one or more reactions relating them, and an equilibrium position from which the unit is displaced. Such a system is a one-way irreversible system with no constraints on symmetry breaking. These systems do not give rise to repeating, self-limited units such as developing organisms. The occurrence of various types of feedback systems provide rate-limiting aspects on the asymmetrical production of one or another of the products. These also provide thresholds the symmetry-breaking system cannot exceed without producing a fluctuation from one steady state to another, a *bifurcation* resulting in a new context (a new morphogenetic step) for the various reactions such that new "morphogens" or new rates of production by persistent "morphogens" are thermodynamically favored. In this manner, morphogenesis is a rate-controlled irreversible sequence of fluctuations from one ordered state to another.

The more realistic models of morphogenesis include some limitations on the growth of morphogen concentration. The mathematical form of these limitation terms is nonlinear for the separate products. Harrison (1982) characterized three different types of resulting models thusly:

Prigogine's Brusselator	Adaptable
Gierer-Meinhardt	Headstrong
Harrison-Lacalli	Crestfallen

I mean that the Brusselator model forms patterns which easily change; peaks will appear, disappear, and move from place to place rather readily. By contrast the Gierer-Meinhardt model is strongly stabilizing for particular features, such as the "head" formation in Hydra and some things that happen in insects. It is likely to be relevant where experiment shows that the character of particular regions persists in grafting experiments. My model differs from both of the preceding. Peaks do not shift easily, but they are not strongly stabilized either, in that they readily split in half by forming a dip at the previous crest. This might have some relevance to branching processes in plants.

However, some models may be linear and still realistic if the rate of bifurcation is high enough that each pattern grows and decays before the next shows up. In such a model, no pattern in the morphogenetic sequence ever reaches such an amplitude that one would require a nonlinear model; such approaches as Turing's (1952) model would then suffice. An example of this latter approach is Kauffman's (1977) modeling of the formation of clone compartment boundaries on a *Drosophila* wing disk. He modeled the disk by an elliptical boundary, with a two-dimensional wave pattern growing inside it; the nodal point for the waves (X, Y) were at $X = Y = 0$. As the simulated disk expanded, patterns grew and decayed in a certain order mimicking that found empirically.

The mechanism Harrison (1981, 1982) and Harrison and Lacalli (1978) favor is that of reaction/diffusion or, as Harrison (1982) put it, "short-range activation; long-range inhibition." This generalization is relevant both to maintenance of a steady state and fluctuations from one steady state to another. In the first case, short-range activation refers to the relatively fast rate of reactions between substrate and enzymes, and long-range inhibition refers to the relatively slow rate of diffusion of product away from the active sites, allowing more reactions to proceed. In the second case, short-range activation refers to continuing asymmetries even in the face of inhibition, and long-range inhibition refers to the boundary conditions limiting the growth of concentrations. In the first case, the result is relative homeostasis; in the second, a bifurcation. Harrison (1982) has shown that differences in initial configurations of active sites can lead to either homeostatic mixtures of products or to segregation of alternate products (a bifurcation of the simplest type) simply on the basis of reaction/diffusion kinetics. Prigogine's Brusselator equation operates on the basis of differential concentrations of various molecules to produce similar results.

The Genome as an Organized Unit

An important aspect of our theory is the idea that the entire developmental program is the causal unit of evolutionary change. If this idea is correct, two predictions may be drawn. First, there should be portions of the genome that act to maintain the unity, or cohesion, of the developmental program. From a developmental (genetic *and* epigenetic) perspective, this would be canalized information. From a purely genetical perspective, canalized information includes regulatory genes and genetic regulatory networks. Second, successful changes in any part of the developmental program are those which integrate with the unchanged parts of the developmental program. This leads to our principle of compensatory changes. Thus, we require that the same dynamics be responsible both for organization into a whole and for production of novelties.

We begin by asking what evidence exists for intrinsic (genic) regulation of development. The first explanation for such regulation was presented by Jacob and Monod (1961), who demonstrated the presence of functional self-regulating genetic units in prokaryotes. Eukaryote regulatory mechanisms are apparently more complicated and several models of gene regulation in ontogeny have been devised (e.g., Britten and Davidson 1969; Zuckerkandl 1976; see review in Raff and Kaufman 1983). These models share common features, including regulatory switch genes and genic control of timing in ontogeny. Regulatory switch genes are those for which a mutation would produce a switch to an alternate developmental pathway during ontogeny; this would produce discontinuous variation in epiphenotypes within a population. Genic control of developmental timing would produce heterochronic changes as different portions of a developmental program proceeded at different rates. For example, if an organism developed external adult characteristics more slowly than normal for its species but matured sexually at the same rate as other members of its species, that organism would become neotenic. Regulatory behavior of portions of genomes, both prokaryote and eukaryote, seems well established. More important to our thesis, however, is the question of whether or not such regulatory components can be responsible for the organization and orderly transformation of form manifested in development.

Much work has been done in this area, most notably by Kauffman (1969, 1971, 1973, 1974, 1981, 1983) and Kauffman, Shymko, and Trabert (1978). According to Kauffman (1983), the study of developmental constraints has, for the past fifty years or so, been associated with the search for the relationship between development and evolution. He identified three areas of analysis that have figured most prominently in such work: (1) allometry, (2) heterochrony, and (3) quantitative genetics. While acknowledging that these three areas of research have gained much insight into the design of living organisms, they do not provide much in the way of causal explanation for either a general mechanism producing the diversity we see or for the possible variety we do not see (i.e., excluded developmental pathways). Kauffman then identified two newer fields of endeavor as potentially contributing answers to many aspects of the contemporary questions. These areas are the structuralist approach of Goodwin (see Goodwin 1982; Webster and Goodwin 1982) and the ensemble constraints approach to gene regulatory networks pioneered by Kauffman himself.

The structuralist program attempts to explain particular phenotypic classes by providing a simple transformation rule for a discrete spectrum of phenotypes. One example of this sort of phenomenon is phyllotaxis in plants. For example, pinecones comprise series of left-handed and right-handed scale spirals. The number of left-handed rows is related to the number of right-handed rows as members of a particular Fibonacci series. Fibonacci series are sequences of numbers in which any new number is the sum of the previous two, such as:

$$1, 1, 2, 3, 5, 8, 13, 21 \ldots$$

or

$$2, 2, 4, 6, 10, 16, 26 \ldots$$

A traditional "canalization" model based in quantitative genetics might account for the apparently random disjunction between 5-8 and 8-13 pairings seen in pinecones by referring to a 5-13 continuum occurring in arbitrarily spaced and crowded thresholds. This would require a large number of ad hoc assumptions and explanations and would provide no boundaries for predictions of future transformations. These are the evolutionary explanations based only on historical contingencies to which structuralists such as Goodwin (1982) have objected so much. Historical contingencies are not the same as historical constraints, however. In rejecting a historical component completely, such an approach risks being unable to discriminate between states that cannot be realized and those not realized because history took an alternative pathway. Such ahistorical structuralism cannot fully accommodate the idea that ontogeny is a nonequilibrium phenomenon requiring a description of past states for its full explanation. A structuralist explanation (see Mitcheson 1977) suggests that all one needs to explain the Fibonacci series in pinecones is close packing of scale primordia on a conical meristem initially, followed by diffusion of inhibitors and expansion of tissue. Mitcheson further suggested that changes in primordia size relative to the meristem, by alterations of diffusion constants or tissue growth rates (see discussion of Harrison's work in the previous section), could produce transformations from one Fibonacci series to another. This explanation, aside from not requiring a set of ad hoc assumptions, predicts a limited set of future transformations and identifies a set of excluded patterns. Another attempt to provide similar structuralist explanations for form regularities is that of Harris and Erikson (1980), who postulated discrete alternative states with restricted paths of continuous and discontinuous transitions between close packing arrangements for biological structures consisting of tubular packings of subunits. In a similar vein, Brooks (1982) drew on common architectural constraints in holdfast (scolex) morphology between two species of cestodes (tapeworms) belonging to the same genus to postulate a discrete set of holdfast structures that would be diagnostic for species not yet discovered. D'Arcy Thompson (1942) and Raup (1968) have also been interested in such phenomena.

Kauffman (1973) and Kauffman et al. (1978) applied this sort of approach to *Drosophila* pattern formation. Called the *combinatorial code–sequential compartmentalization* model, Kauffman's theory is explicitly structuralist. For a given set of initial conditions, this theory postulates a set of "code words" corresponding to a finite set of alternative developmental decisions at a particular point in ontogeny, and it postulates a constrained set of transformations among them. This model allowed Kauffman to predict a large number of specific classes of transformations seen in homeotic mutants of *Drosophila*. Kauffman (1983) asserted that such a combinatorial character in developmental programs reflects a very general principal of genomic dynamical organization, which acts as an intrinsic constraint in evolution. Kauffman et al. (1978) produced a model, based on reaction/diffusion gradients, to account for position gradients in the embryo-triggering sequential developmental commitments. Similarly, Murray (1981) and Bard (1981) proposed a reaction/diffusion model to account for stripe and patch coat color markings on a variety of mammals.

Kauffman (1983) finished his discussion of the structuralist approach thus:

> In short, when we genuinely understand the control of shapes, patterns, morphologies, at least some evolutionary transitions between a restricted family of genuinely neighboring forms should fall out as predictions, not mere arbitrary empirical correlations. In this sense, the concept of "devel-

opmental constraints" connotes the relative ease of transformation in evolution to a *delineated* set of neighboring organisms. Such constraints are internal factors in evolution channeling the progression of forms.

This suggests to us that Kauffman does not believe there is a contradiction between the structuralist approach and evolutionary history, that is, a hierarchical set of relationships. Nor does it appear that Kauffman sees any contradiction between the structuralist approach and the kinetic (reaction/diffusion) approach. Rather he seems to regard canalization as an outcome of certain evolutionary processes rather than an active process itself, in contradiction to the ideas of many quantitative geneticists (e.g., Waddington 1977). We agree.

The second new approach to studying developmental constraints is that of *ensemble constraints* (Kauffman 1983). The basis of this concept derives from the following excerpt from Kauffman:

> I turn now to a different sense of "developmental constraint" which may play a critical role in evolution and has received little attention. The detailed mechanisms of gene regulation in eukaryotes are not yet known. Points of regulation include transcription to heterogeneous nuclear RNA, splicing, capping, polyadenylation and transport of mature messenger RNA to the cytoplasm, initiation and termination of translation, and post-translational modifications. . . . Among these, it seems clear that the early expectation of genetic cisacting sites and transacting sites is substantiated. Families of such genetic loci have now been found in yeast, mouse, *Drosophila,* and maize. . . .
> Furthermore, it is now widely appreciated that tandem duplications have arisen in evolution . . . while recent evidence demonstrates that fairly rapid dispersion of some genetic elements occurs through chromosomal mutations . . . including transpositions, translocations, inversions and recombination. . . . Dispersal of loci by processes such as these provide the potential to move cis acting sites to new positions and thereby create novel regulatory connections, opening novel evolutionary possibilities. . . . *This raises a general question: If duplication and dispersion occur with* characterizable probabilities per site, *then in the absence of selection, is it possible to build a statistical theory of the expected control structure of a genetic regulatory system after many such transformations?*

Kauffman investigated this possibility by modeling the effects of locus duplication and dispersion on simple regulatory gene networks; dispersion was modeled by the use of transposition alone, rather than attempting to include such factors as recombination, inversion, deletion, translocation, and point mutations. The composition of two types of simulated parts of chromosomes and of their expected regulatory networks is shown in Figs. 3.5 to 3.8. An additional simplifying assumption of the model is that loci cannot be destroyed; thus, the rate of locus formation depends on the number of existing copies, and the growth of each locus is viewed as being stochastic and exponential. This model, then, is certainly highly simplified, but the simplifying aspects make it an incomplete rather than unrealistic model of regulatory gene network dynamics. Kauffman himself stated that the primary purpose of the model is to examine the regulatory architecture after many instances of duplication and transposition. The results of two simulations based on different initial networks in which transposition occurs nine times for every duplication in the course of 2000 iterations are shown in fig. 3.9. Because of the effects of transpositions there is a random spacing of the regulatory connections throughout the genome. Therefore, once a sufficient number of iterations has been performed, the overall architectures of all systems will have a fair degree of similarity. The surprising result of simulations with this model, according to Kauffman, is that the randomized genetic networks exhibit spontaneous highly constrained, ordered dynamic properties. For example, 60% to 70% of the genes in any simulation emerge from the simulation in fixed "on" or "off" functional categories. That 60% to 70% constitutes a large subgraph, called a *forcing structure,* of the entire network. The forcing structure blocks the propagation of *varying* gene products from genetic loci that are not parts of the forcing structure. The genes that are not part of the forcing structure form small clusters, each of which is *functionally* isolated (if not spatially isolated) from influencing other such clusters by the constraints of the forcing structures. Each of the subsystems has only a small number of discrete modes of behavior available to it; these may be in the form of alternate steady states or oscillating patterns of gene activity. In short, the networks show that the regulatory genes, even just the *cis-* and *trans*-acting ones, tend to exhibit dynamic order involving the entire genome. Their inherent behavior is combinatorial; that is, the different behavioral modes exhibited by, say, the differentiated cell types in an organism, comprises the possible combinations of the

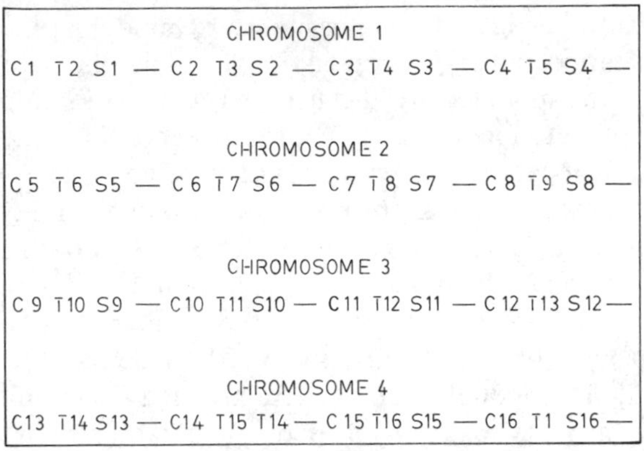

```
                        CHROMOSOME 1
C1  T2 S1 — C2  T3 S2 — C3  T4 S3  — C4  T5 S4 —


                        CHROMOSOME 2
C5  T6 S5 — C6  T7 S6 — C7  T8 S7  — C8  T9 S8 —


                        CHROMOSOME 3
C9  T10 S9 — C10 T11 S10 — C11 T12 S11 — C12 T13 S12 —


                        CHROMOSOME 4
C13 T14 S13 — C14 T15 T14 — C15 T16 S15 — C16 T1 S16 —
```

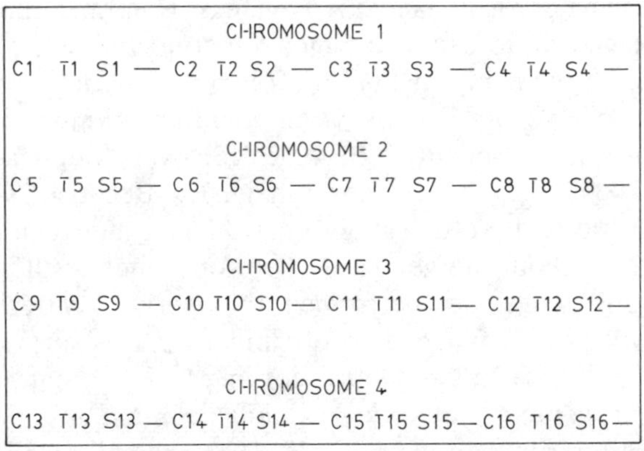

```
                        CHROMOSOME 1
C1  T1 S1 — C2  T2 S2 — C3  T3  S3  — C4  T4 S4 —


                        CHROMOSOME 2
C5  T5 S5 — C6  T6 S6 — C7  T7  S7 — C8  T8 S8 —


                        CHROMOSOME 3
C9  T9 S9  — C10 T10 S10 — C11 T11 S11 — C12 T12 S12 —


                        CHROMOSOME 4
C13 T13 S13 — C14 T14 S14 — C15 T15 S15 — C16 T16 S16 —
```

Figs. 3.5 to 3.8. Hypothetical chromosomes and gene actions for regulatory ensemble simulation. (Modified from Kauffman 1983.)

Fig. 3.5. The genome is assumed to have only four kinds of genetic elements, *cis* acting (*Cx*), *trans* acting (*Tx*), structural (*Sx*), and empty (−). All types of elements are indexed. Any *cis*-acting site is assumed to act in polar fashion on all *trans*-acting and structural genes in a domain extending to its right to the first blank locus. Each indexed *trans*-acting gene, *Tx*, regulates all copies of the corresponding *cis*-acting gene, *Cx*, wherever they may exist on the chromosome set. Structural genes, *Sx*, play no regulatory roles.

Fig. 3.6. Sixteen sets of triads of *cis*-acting, *trans*-acting, and structural genes separated by blanks on four "chromosomes" are arrayed.

functionally isolated subsystems that are allowed by the forcing structure.

Kauffman also analyzed four aspects of networks in which a given number of genes (N) were linked at random by a given number of regulatory connections (M): (1) the number of genes directly or indirectly influenced by each single gene, that is, the descendants

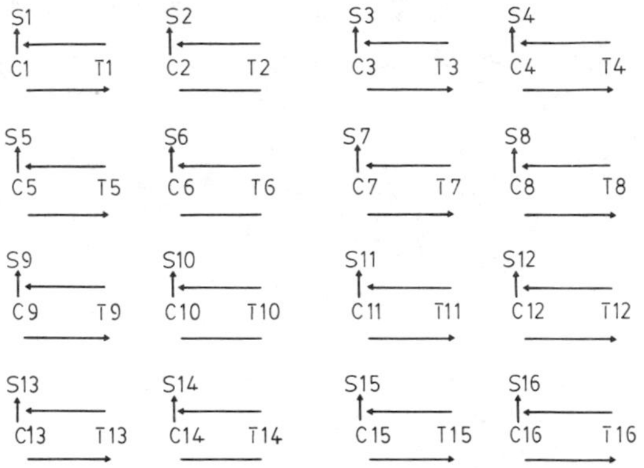

S1   S2   S3   S4
C1   T1   C2   T2   C3   T3   C4   T4

S5   S6   S7   S8
C5   T5   C6   T6   C7   T7   C8   T8

S9   S10   S11   S12
C9   T9   C10   T10   C11   T11   C12   T12

S13   S14   S15   S16
C13   T13   C14   T14   C15   T15   C16   T16

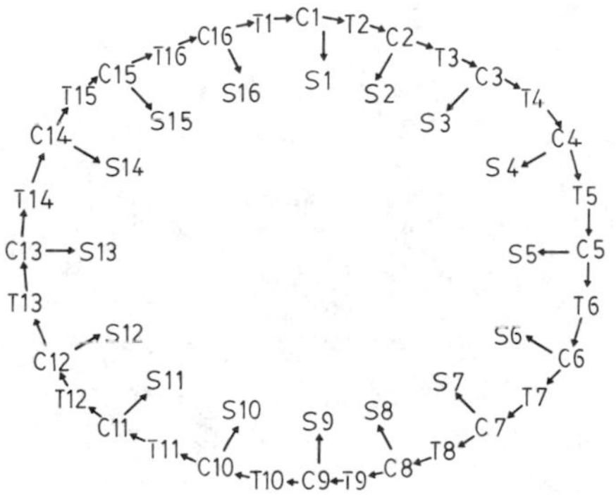

[Fig. 3.8 circular diagram: a single long feedback loop running T1→C1→T2→C2→T3→C3→T4→C4→T5→C5→T6→C6→T7→C7→T8→C8→T9→C9→T10→C10→T11→C11→T12→C12→T13→C13→T14→C14→T15→C15→T16→C16, with S1, S2, S3, S4, S5, S6, S7, S8, S9, S10, S11, S12, S13, S14, S15, S16 arrayed around the ring]

Fig. 3.7. A graphic representation of the control interactions among these hypothetical genes is shown in which an arrow is directed from each labeled gene to each gene that it affects. Fig. 3.7. *C1* sends an arrow to *T1* and an arrow to *S1*, while *T1* sends an arrow to *C1*. A similar simple architecture occurs for each of the sixteen triads of genes, creating sixteen separate genetic feedback loops. By contrast, in fig. 3.6, each triad carries the indexes (*Cx, Tx + 1, Sx*), while the sixteenth is *C16, T1, S16*.

Fig. 3.8. This permutation yields a control architecture containing one long feedback loop. (Redrawn from Kauffman 1983.)

from each gene; (2) the *radius* from each gene, that is, the minimum number of steps for influence to propagate to all of its descendants; (3) the fraction of genes lying on genetic feedback loops among the total number of genes; and (4) the length of the smallest feedback loops for any gene on a feedback loop. The results, based on $N = 200$ genes and $M = 0 - 720$ regulatory connections (figs. 3.10 to 3.13) indicate that these four properties depend both on the total number of genes and on the number of regulatory connections among them. As N increases past M, large forcing structures begin crystallizing in descendants; some genes begin influencing a large number

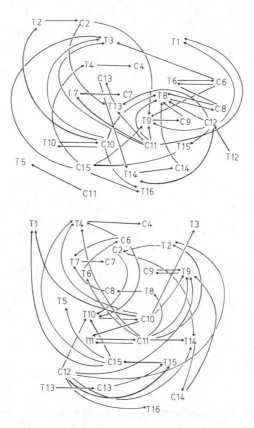

Fig. 3.9. Hypothetical regulatory gene ensembles generated by simulation model. Directions of arrows indicate direction of regulatory influence. (Modified from Kauffman 1983.)

of other genes. By the time $M = 3N$, any gene may directly or indirectly affect up to 87% of the total genome. The relationship between the number of descendant genes and M is shown in fig. 3.10 In fig. 3.11, the mean radius as a function of M is depicted. The values are low when there are few connections, that is, each gene can influence all of its descendants in a few steps. The values become maximal when M is between $1.5N$ and $2.0N$; each gene can influence a large number of other genes because the network is still relatively sparse. As the network becomes more complex, the values begin to decrease as newer, shorter connections become possible. As the system approaches the total possible number of connections, we might expect the mean radius M value to asymptotically approach a steady state. Third, the fraction of genes lying on feedback loops

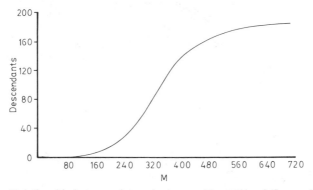

Fig. 3.10. Relationship between descendant genes (the number influenced by genes activated earlier in ontogeny) and the number of regulatory connections (M) in network ensemble simulations. (Modified from Kauffman 1983.)

parallels the mean number of genes descended from each gene (fig. 3.12)—as M increases so does the chance of forming closed loops. And, last, the lengths of the smallest loops on which genes lie parallels the mean radius distribution, reaching a maximum when M reaches $1.5N$ to $2N$, and then declining as M increases (fig. 3.13).

Kauffman's model allows us to envision a unified mechanism governing the expected *architecture* of regulatory gene networks affected by a wide variety of genomic changes. Once realized, such a theory would allow us to test predicted dynamic behavior of regulatory networks. Kauffman (1983) concluded his discussion of the ensemble theory as follows: "insofar as duplication and dispersion of genetic loci throughout the genome tend toward a particular statistically robust class of regulatory architectures, their typical struc-

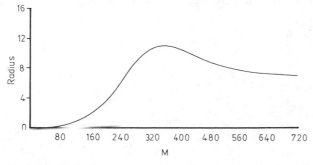

Fig. 3.11. Relationship between mean radius (the minimum number of steps necessary to propagate gene influence from one gene to its descendants) and number of regulatory connections (M) in network ensemble simulation. Note self-stabilizing or self-organizing influence of past connections. (Modified from Kauffman 1983.)

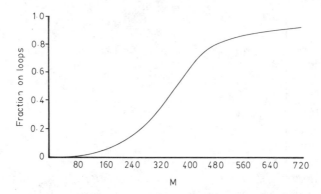

Fig. 3.12.   Relationship between fraction of genes lying on closed, or feedback, loops and number of regulatory connections (M) through time in network ensemble simulations. As M increases, so does the number of closed loops. (Modified from Kauffman 1983).

tural and typical dynamical features constitute developmental constraints in evolution which may be very powerful."

Of course, the "randomizing" process in real genetic regulatory systems would be mutations in the regulatory programs. If, as Kauffman suggested in the excerpt above, regulatory gene networks are a very powerful constraining factor in evolution, changes in the regulatory networks themselves should be powerful promoters of evolutionary change, albeit along constrained pathways. Therefore, it seems reasonable to postulate classes of changes and classes of expected outcomes of regulatory gene, or gene network, mutations.

Following the discussion by Hollinger and Zenzen (1982; see also chapter 2), we would expect genomes to exhibit ensemble behavior if they are nonequilibrium systems. Self-regulation would be an ex-

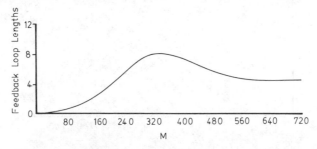

Fig. 3.13.   Relationship between lengths of smallest genetic feedback loops and number of regulatory connections (M) through time in network ensemble simulations. Note self-stabilizing or self-organizing influence of past connections. (Modified from Kauffman 1983.)

pected emergent property of such systems because the requirement that new microstates be accessed by the ensemble. This inevitably leads to increasing specialization determined at least in part by the microstates already accessible to the ensemble. The more complex the ontogeny, the greater the degree of self-regulation, and yet the developing ensemble must access new microstates in compliance with the second law. The result is evolution of increasingly complex, self-regulated ontogenies through time. The only way to avoid this would be to turn on all genes simultaneously and ensure that each gene exerted an equal influence on the ensemble. The result would be a perfect, immutable sphere. This is a major clue to the problems encountered previously in attempting to understand coherent genome behavior—"canalization" is not a process, but an inevitable consequence of the dynamics of ontogeny.

Zuckerkandl (1976), whose views parallel those of Kauffman, stated:

How about mutations that affect the topological relationships between controller genes and structural genes? A reshuffling of regulatory circuitry, if all theoretically possible combinations . . . were considered, would again lead to a relatively limited number of, this time, consistently discontinuous phenotypic expressions. A certain limitation of the variety of possible phenotypic effects due to alterations in gene regulation would again be expected.

He continued:

The intrinsic component [of evolution] should be linked to the characteristics of phenotypic effects of mutation in controller genes.

Finally:

Many structural and functional developments will not occur at all, because they are incompatible with the existing genetic constellation.

Zuckerkandl suggested that mutations in regulatory genes would affect four basic areas of gene function: (1) the time in relation to a developmental stage at which a gene product is made; (2) the duration of its period of synthesis; (3) the rate of its synthesis; and (4) the total output of gene product. At a higher macroscopic level, that of development, he suggested two classes of changes: those in which the topology of the regulatory network remained intact and those in which it was changed. For mutations in which the regulatory topology remains intact, expected changes would be manifested in changes in time of onset and rate of transcriptional activities and/or in rate of translational activities. Mutations leading to a change in regulatory network topology would be expected to affect the temporal and/or spatial context of transcriptional activity. According to Zuckerkandl, mutations that do not change the regulatory topology would account for the control of continuous variation, whereas mutations altering the topology lead to discontinuous variation. This is so because changes in a spatiotemporal context can cause inducers and competent tissue to fail to coincide, thereby preventing, rather than slowing down or speeding up, the emergence of a structure. Because only processes leading to discontinuities result in evolutionary change, Zuckerkandl viewed changes in developmental programs (i.e., the largest functional subunits of the ontogenetic program of the organism) as being the focal point of discussion about evolutionary "advancement."

Both Zuckerkandl (1976) and O'Grady (1985) found it useful to think of changes in gene programs in development in terms of changes in life cycle stages (*semaphoronts* of Hennig 1966). Although Zuckerkandl was interested in gene action and O'Grady in phenotypic expression, their classes of changes overlap a great deal. O'Grady recognized six classes of changes in ontogenetic sequences associated with evolutionary divergence (cladogenesis). They are terminal addition, terminal substitution, terminal deletion, nonterminal addition, nonterminal substitution, and nonterminal deletion. We will consider all six classes.

Terminal Changes

Terminal addition. There is close agreement between Zuckerkandl's and O'Grady's views. Zuckerkandl (1976) wrote: "Processes of terminal program additions imply that contemporary larval stages should be closely related to ancestral adult stages (such as the trochophore larva of annelids to the adult ancestors of ctenophors . . .) and present adult stages, strictly speaking, should have no equivalent in the distant past." O'Grady (1985) notes: "Terminal addition, therefore, allows application of the hypothesis of recapitulation in phylogenetic reconstruction."

As an example of terminal addition, O'Grady chose mammalian hair. Oster and Alberch (1982) have pointed out that hair develops in mammals as a result of an epithelial invagination added to the epithelial evagination producing scale precursors. Distribution of hairs is determined by scale distribution. Mammals showing evidence of scalation (rats' tails, scaly anteaters, armadillos) exhibit a pattern of three hairs projecting from underneath each scale; hairs are arranged in triplets even in mammals showing no evidence of adult scalation.

Terminal substitution. As an example of this class of change, O'Grady used the substitution of feathers for scales in birds. The epithelial evagination producing scales in ancestral archosaurs produces feathers in birds. Scales may be present on birds' legs, but on the other parts of the body, only feathers appear. Even though birds such as owls, ptarmigans, and domestic fowl have feathers on their feet that appear to be feathers growing from scales, experimental studies have shown that the scales erupt from underneath the feather (Senger 1976; Senger and Pautou 1969).

Terminal deletion. An illustration of regulatory gene program deletion is limb loss in salamanders, snakes, and lizards. Zuckerkandl suggested that the structural genes necessary for limb production are probably still intact, but function for other purposes because (1) there are probably no "bone structural genes" or "nerve structural genes" specially designated for limb morphogenesis and (2) there are probably no "limb structural genes" per se. Therefore, it is the presence or absence of a particular regulatory network that determines that a cluster of structural genes will act to produce limbs. Zuckerkandl also suggested that terminal deletion might be accompanied by terminal substitution. We will discuss such correlated changes in the next section. O'Grady's example of terminal deletion involves the loss of teeth in birds. In this case, a terminal ontogenetic product has not formed. Because tooth enamel can be induced from chick epithelium (Kollar and Fisher 1980), it appears that the necessary structural genes are intact (even after more than 100 million years) and that the evolutionary change has involved an alteration in a regulatory program.

Nonterminal Changes

In general, if ontogeny is a causal sequence and if evolutionary divergence is largely a result of changes in regulatory programs, then nonterminal changes in ontogeny should be more difficult to achieve than terminal ones. Because of the interconnections among gene activities, many changes in nonterminal stages will not be realized because they interfere with regulatory processes associated with the rest of the ontogenetic program. We might also expect that the earlier in ontogeny such interference occurs, the less the chance it will successfully integrate with the rest of the sequence. Alternatively, such alterations, when successful, should have profound effects on the species involved.

Nonterminal addition. One of the most familiar examples of nonterminal addition is that of the intercalation of a placenta in the ontogeny of mammals. Although this could be viewed mechanically as a relatively minor nonterminal ontogenetic change, it had profound physiological, ecological, and behavioral consequences. In contrast to terminal addition, in which new adult features have no counterparts in ancestral forms, nonterminal addition results in new intermediate developmental stages (larvae, pupae, juveniles, etc.) or features that have no counterparts in ancestral forms.

The most significant manifestation of nonterminal additions is that of the intercalation of larval stages creating complex life cycles (e.g., the digenetic trematodes) and the phenomenon of metamorphosis. Two processes must operate to produce such ontogenetic programs. First, there must be successful nonterminal additions. This would lengthen the developmental sequence and affect the subsequent stages in the developmental sequence. If, in addition, some or all the subsequent changes are controlled by independent regulatory networks, the intercalation of a nonterminal addition may effectively render the later (subsequent) regulatory gene programs a "package" that can be triggered by the termination of activity or by a specific product of the larval stage immediately preceding it. In holometabolic (completely metamorphosing) insects and in salamanders, the adult "package" is triggered by the hormones ecdyson and thyroxin, respectively, produced by the larval stages preceding it. If a similar packaging phenomenon occurs in nonterminal portions of the ontogeny, each of the packages may be referred to as a backbone program; Zuckerkandl (1976) stated that any discontinuation of one backbone program concomitant with the activation of another represents a case of metamorphosis. The mere intercalation of a nonterminal addition may retard the onset of initiation of the terminal stages, rendering them effectively a package even in the absence of higher level regulatory constraints. However, evidence of higher level constraints is known. For example, in holometabolic insects, the programs leading to adult structure are activated in early development but shut down and are held in imaginal disks when larval stage development is initiated; then they are reactivated once larval development has terminated. For such ontogenetic organization, one would expect that the larval and adult programs could be highly independent in terms of relative degree of relative differentiation. For example, Wigglesworth (1954) reported that adults of two species of butterflies, *Acronyta psi* and *A. tridens,* were virtually indistinguishable but their larvae were very dissimilar.

Nonterminal substitution and nonterminal deletion. Empirical evidence for independent episodes of these phenomena is scanty. O'Grady mentioned, as an example of nonterminal substitution, the reversal of the orientation of the appendicular to axial skeleton in turtles relative to other tetrapods. A possible example of nonterminal deletion involves the formation of a reduced fibula, relative to other archosaurs, in birds.

Entropic Behavior of Ontogenesis

The previous section provided evidence that during development the entire genome acts as a unit, controlled by its regulatory gene network. The effects of such a system are ontogenetic sequences that begin as relativly simple organized structures and end as highly complex organized structures. The dynamics of these transformations indicate a strongly deterministic tendency toward that endpoint of ontogeny. This tendency toward highly constrained regulatory architecture is not lost even if the regulatory connections are randomized. However, because ontogenesis is a process whereby the information (instructions) specifying the organization and transformation of the organism through time is contained in the zygote, the boundary conditions for each episode of ontogenesis are present from the start. This means that ontogenesis is an entropic phenomenon in which the increase in structural complexity observed is a manifestation of the actual expression of information contained as potential microstates in the embryo, that is, the progressive *accession* of microstates comprising the regulatory network. The rate and degree of expression are constrained by the buildup of structural complexity, including homeostatic mechanisms, which provide for the persistence of an organism during ontogenesis and after it is completed.

We may document the entropic behavior of structural change in ontogeny by measuring the statistical entropy of regulatory gene

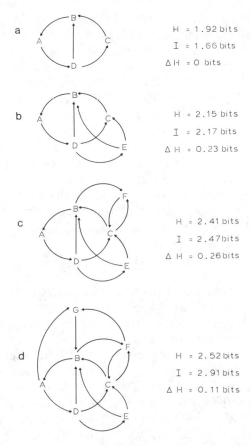

Fig. 3.14. Schematic representation of regulatory gene ensembles at four sequential stages in the ontogeny of a cell. H = Entropy; I = information; ΔH = change in entropy from one stage to the next.

ensembles. Four such ensembles, representing an organism during four consecutive stages in the ontogeny of a specialized cell type are depicted in fig. 3.14a–d. In fig. 3.14a, genes A, B, and C each exert regulatory effects on one other gene, whereas D regulates two genes. There are five regulatory connections, in total, so the probability associated with A, B, and C is 1/5, and the probability associated with D is 2/5. The entropy of the ensemble is thus:

$$H_a = -(3)(1/5 \log_2 1/5) - (2/5 \log_2 2/5)$$
$$= 1.92 \text{ bits}$$

By the same calculation, the probabilities associated with fig. 3.14b are 1/8 for A, B, and C; 2/8 for E; and 3/8 for D. The entropy of this ensemble is

$$H_b = -(3)(1/8 \log_2 1/8) - (2/8 \log_2 2/8) - (3/8 \log_2 3/8) = 2.15 \text{ bits}$$

For fig.3.14c, the probabilities are 1/11 for A and F; 2/11 for B, C, and E; and 3/11 for D. The entropy of this ensemble is

$$H_c = -(2)(1/11 \log_2 1/11) - (3)(2/11 \log_2 2/11) - (3/11 \log_2 3/11)$$
$$= 2.41 \text{ bits}$$

In a similar manner: $H_d = 2.52$ bits

This entropy measure holds even if some of the genes considered were active in fig. 3.14a and inactive in fig. 3.14d, because the stage shown in fig. 3.14d could not have been achieved without passing through the first three stages. In this sense, all genes active in early stages are a part of the entropy-producing ensemble at all subsequent stages. In this way, it is easy to see that the sequential activation of portions of the genome during ontogeny represents an entropy-producing accession of more and more microstates into a growing ensemble in accordance with the requirements of nonequilibrium systems.

Multicellular organisms are aggregates of differentiated cells related to each other by a common cell lineage. This can be represented symbolically by a directed graph (e.g., fig. 3.15). Each of the cells could be represented by a regulatory gene network such as those shown in the previous fig. 3.14 and in fig. 3.9. It is possible to estimate the statistical entropy of the lineage of cells treated as an ensemble containing various numbers of the three different cell types. Because we have moved up a functional level from regulatory gene networks to cell lineages, the entropy of the single-cell stage is zero. This indicates that no transformation of the system has occurred at this functional level. For the two-cell stage, there are two different cell types, each of which is derived from the same parental cell. In calculating the entropy of the two-cell stage, we must treat it as a three-cell ensemble, with the parental cell counted once for each descendant cell. Probabilities are thus 1/4 for each descendant cell and 1/2 for the parental cell. Further, since one of the descendant cells is the same cell type as the parent, the entropy is calculated as

$$H_2 = -(3/4 \log_2 3/4) - (1/4 \log_2 1/4)$$
$$= 0.81 \text{ bit}$$

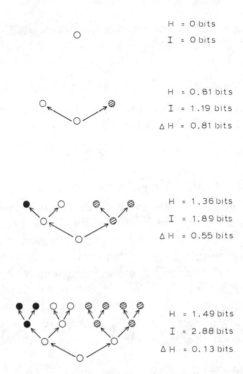

Fig. 3.15. Schematic representation of a cell lineage shown at four stages as directed graphs. H = Entropy; I = information; ΔH = changes in entropy from one stage to the next.

where 3/4 is the probability value for the parental cell type and 1/4 is the value for the derivative cell type. Calculations for the four-cell and eight-cell stages, following the rationale stated above (counting parental cells twice in calculating probabilities), are as follows:

$$H_4 = -(1/10 \log_2 1/10) - (5/10 \log_2 5/10) - (4/10 \log_2 4/10)$$
$$= 1.36 \text{ bits}$$

and

$$H_8 = -(4/22 \log_2 4/22) - (8/22 \log_2 8/22) - (10/22 \log_2 10/22)$$
$$= 1.49 \text{ bits}$$

Even though the parental cell does not exist at the same time as the cells derived from it, it remains historically as one of the accessed microstates of the system, just as regulatory genes that may be inactivated remain historically part of the regulatory network. It is evident from the entropy values that some resolution of the total entropy increase in the system is lost when moving from one functional level to another. Fortunately, it is not absolute magnitude of entropy production that is important to our considerations, but relative entropic behavior.

For example, despite the problem of relative levels of functionality, the entropic behaviors of developing cells and of developing cell lineages are identical in two fundamental respects. First, during the developmental process, the statistical entropy of the system increases at both functional levels. Second, the amount of entropy increase from one stage to another through time always decreases. This decreasing relative rate of entropy increase results from two phenomena. First, historically accessed microstates must be included in a complete description of a developing system. This is a direct indication that history constrains entropy increase, in accordance with generalized predictions of the behavior of nonequilibrium entropy systems presented in chapter 2. Second, as a developing cell type or organism becomes more complex, accession of new microstates affects a progressively smaller proportion of the system. Thus, the actual increase in entropy lags further behind the maximum possible entropy increase. In a manner analogous to the dynamic discussed in chapter 2, this increasing lag signals increasing organization, while the increasing entropy signals increasing complexity. For example, the organization values ($\log A - H$) for the ensembles shown in fig. 3.14a–d are 1.66, 2.17, 2.47, and 2.91 bits, respectively; for those in fig. 3.15a–d, the values are 0, 1.19, 1.89, and 2.88 bits, respectively.

One mechanism producing such macroscopic constraints on a developing ensemble of cells is the production of cell-adhesion molecules (CAMs) (Edelman 1984a, 1984b). Edelman (1984a) asserted:

> Development is historical. Not only primary induction but also the entire sequence of secondary inductions . . . depend on the apposition of particular cells that have had different histories. This extraordinary sequence of events takes place in a well-defined temporal sequence and is organized in stages, with structures in each stage serving as the basis of structures in succeeding stages.

and

> There are two alternative ways patterns might be formed at the cellular level without the direct intervention of some kind of "little architect" or "construction demon."

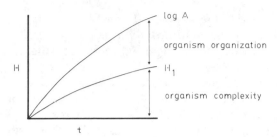

Fig. 3.16. Summary schematic representation showing ontogeny as a process of increasing organization and increasing complexity through time.

One of the two mechanisms Edelman suggested and rejected was the geometric building-block approach also criticized and rejected by Harrison (1981, 1982). The other was the dynamic production of molecules mediating cell-to-cell adhesion during development. These would act to constrain cell movements and changes in cell size and shape. This in turn would maintain asymmetrical energy dissipation (symmetry breaking), resulting in the emergence of organized patterns. Or, as Edelman (1984a) explained:

> In this kinetic, far-from-equilibrium situation, pattern results from the play of energy as it is dissipated.

Clearly, Edelman's view of ontogeny is that of a nonequilibrium entropic phenomenon in which initial conditions (history) play an important role. The effect of some of those initial conditions, specifically of the CAMs, is to constrain entropy production, resulting in organization. This is entirely consistent with our views. In fig. 3.16, the historically constrained entropic emergence of organized and complex form in ontogeny is summarized.

These aspects of ontogeny show organized increases in complexity. Furthermore, the increase exhibits a trend toward minimizing entropy increases because of the historical constraints placed on complexity increases. Thus, coherent structural change of the type suggested by Ho and Saunders (1981; see discussion in chapter 1) is an emergent property of ontogenesis as a nonequilibrium phenomenon. This trend will occur regardless of the effect particular environmental regimes might have on particular evolving lineages. It is not necessary to postulate a direct influence of the environment or to invoke negentropic processes.

The "negentropic" behavior associated with the decreased rate of thermodynamic entropy production during ontogeny is not a manifestation of a negentropic process. Rather, it is an indication that energy flows are not the determining factor in the dynamics of ontogeny, but only a necessary boundary condition. Information is the determining factor, and the progressive expression of information during ontogeny is accompanied by an increase in the statistical entropy of the ensemble representing the developing organism. Because ontogeny is a nonequilibrium phenomenon, as an organism approaches maximum complexity its rate of entropy increase declines, and as that happens, the rate of energy uptake and dissipation decreases. The principle of minimum entropy production holds true in either case, as Zotin and Zotina (1978) suggested on purely energetic grounds.

By associating macroscopic information with developmental causality, we are not calling on inherent biological information to act as if it were a computer program specifying the step-by-step construction of an organism. Rather, we expect the second law to produce increasing complexity in developing organisms and require that developmental programs only be analogous to computer algorithms that specify constraints on the extent to which such entropy-driven complexity can increase. The macroscopic information, or organization, is an assessment of the self-organizing influence of the algorithm. The amount of information required to specify an algorithm placing specific constraints on increases in complexity may be much less than needed to specify the step-by-step construction sequence.

The source of developmental algorithms is reproduction; hence, any algorithm is part of the initial conditions for a developing organism. Since the amount of information needed to specify an algorithm may be much less than that required to specify a construction sequence, the amount needed to transform one algorithm into another may be very small indeed. Production of new algorithms increases the complexity of algorithms that have evolved and is thus an entropic phenomenon. This transformation of developmental algorithms will be a nonequilibrium phenomenon if (1) replication rates are higher than mutation rates in reproduction, (2) at least some variations in algorithms produce the same ontogenetic end product (e.g., homozygous dominant and heterozygous individuals for some traits), and (3) not all changes in algorithms produce viable, reproducing organisms. Such "sorting" processes constrain the entropic change in developmental algorithms, ensuring that actual complexity increases will lag behind the maximum possible.

During the course of an individual ontogeny, we find energy dissipation (entropy production) constrained by a historically established developmental algorithm. During the sequence of ontogenies characterizing a reproductive continuum, we find information dissipation (also entropic) as developmental algorithms are transformed. Thus ontogeny, and the sequence of ontogenies produced by reproduction, are characterized by two coupled entropic processes, energy dissipation and information dissipation. Since the information is part of the initial conditions for development of any individual organism constraining energy dissipation pathways, and since the entropic changes in developmental algorithms occur at a slower rate than energy dissipation in each ontogeny, we expect ontogenesis to be primarily determined by information and not by energy.

Compensatory Changes

If an ancestral ontogenetic program is highly organized and determinate, changes in it will cause departures from that determinism. What leads us to believe that such changes by themselves (i.e., without recourse to natural selection) will produce other determinate programs rather than progressively less coherent systems? The answer requires that we reexamine the developmental context in which new programs emerge.

The potentiality for a new ontogenetic program, distinct from the parental one, is produced by the parent itself, in the form of gametes or their equivalents (in asexual species). Therefore, gametes are a part of the developmental program of a parent, whether they contain mutant genetic material or not. Some gametes may be flawed genetically to such an extent that they are incapable of forming a viable zygote with another gamete. The result is a loss of order in the species. Successful fusion of two gametes to form a zygote triggers a progressive differentiation. This differentiation, when measured by structural or physiological change, can be seen to follow the expectations associated with a nonequilibrium entropy system. When measured in terms of gene action (Kauffman 1983), this differentiation is manifested by a proliferation of regulatory gene connections. For successful zygotes, the end product is one which is viable and fertile. Unsuccessful zygotes are either inviable or infertile.

Once a series of new ontogenetic programs has emerged from mutant gametes via development, we might map the temporal sequence of emergence of such programs. The mapping of programs would provide a view of related but discontinuous ontogenies in terms of their final regulatory gene architecture (and phenotype). If we recognize that each ontogenetic program originated as part of a parental ontogenetic program, it becomes apparent that, so long as successful reproduction occurs, the decrease in order and organization at the population level (see chapter 4) caused by new ontogenetic programs is the result of an increase in number and variety of types of organized systems; that is, an increase in the number of kinds of phenotypes. The mutant gametes never fragment the developmental programs of the parents in which they form. Further, because a new complex ontogenetic program emerges from part of a previous program, the production of new complex ontogenetic programs parallels the production of new complex structures in ontogeny. At one level, gametes are a manifestation of the increasing complexity of an organism as it develops; at the next higher level, a gamete is a manifestation of the increasing complexity of the species.

Of course, if the expected state according to our view is continual organization in the face of the changes that mark evolution, natural selection cannot fill the role of organizer. Some authors (e.g., Charlesworth, Lande, and Slatkin 1982; Zuckerkandl 1976) seem to believe that natural selection accounts for the organization seen in ontogenetic programs. But organization is an emergent property of development, and production of gametes is a part of development. Natural selection will never have the opportunity to operate on disorganized units of any kind. Therefore, the effects of natural selection might be seen in terms of departures from, or reinforcement of, expected order or orderly change, not in terms of bringing order from chaos.

The question of the mechanical origin of evolutionary novelties has been handled in an ad hoc manner by evolutionists. This is undoubtedly due in part to the fact that evolutionary biologists are not as much concerned with the *origin* of novelties as they are with the *fate* of novelties. Darwin and Lamarck believed that novelties arose as a de novo response to some adaptive need by the organism vis-à-vis its extrinsic environment. The current view is an improvement: novelties are the result of genetic mutations that arise at random with respect to the extrinsic environment, and only those which are adaptive are favored and persist. If the effects of these mutations are manifested bit by bit through time, the phenomenon is termed *microevolution,* or phyletic gradualism; if many such effects are realized simultaneously, one refers to *macroevolution,* and begins speaking of coadapted gene units (see Mayr in Tax 1959). In any event, the view is quite clearly that variations in developmental

programs are undirected in themselves, but the extrinsic environment acts as a determinate force, molding or trimming ontogenetic
programs to the current adaptive needs.

In our view, the genealogically determined portion of the genome
and not the extrinsic environment is the major deterministic factor
in evolution. Therefore, novelties, if they occur, must arise either
from a restricted part of the genome in each generation, that is, the
portion not involved in the regulatory gene network, or they must
be a restricted subset of all possible mutations, that is, those which
integrate with the rest of the genome. There is no such thing as a
random mutation; they may arise randomly with respect to the extrinsic environment, but they are exceedingly nonrandom with respect to the rest of the genome.

Zuckerkandl (1976) suggested that in the absence of alteration in
gene regulatory structure, the number of possible (= viable) and
distinct phenotypic expressions should be reduced; further, he suggested that even in the event of alterations in the regulatory structure, a limited variety of possible new phenotypes would be expected:

> The striking directionality of evolution . . . is probably due,
> in part, to the intrinsic limitation of viable and morpholog-
> ically [phenotypically] detectable mutants in controller DNA.
> The recurrence of the same morphological changes in in-
> dependent lineages may be considered as far more probable
> than would be the case if the situation were not as described.
> Similar morphological changes may thus spread in several
> related lineages, even when not . . . advantageous but sim-
> ply not unfavorable. Accordingly, the vectors of evolution
> may be in part of intrinsic origin. More generally, it should
> be based on the fact that, in the succession of genomes by
> filiation, the state that follows has its possibilities restricted
> by the state that precedes.
>
> The intrinsic vector of evolution is not, in fact, a line but
> a vectorial zone. This oriental zone is defined by the limited
> possibilities of new "orchestrations" of gene activities,
> starting from the orchestration that exists.

Kauffman (1983) noted that when he deleted "inactive" structural
genes from his model ontogenetic sequences, only a small number
of differentiation steps were perturbed; the same was true for simulations imitating the effects of mutations similar to those manifested
by homeotic mutants. He stated:

most of the mutants in this class affected much the same small differentiation steps in the [probability] matrix, and typically, a given . . . element was more likely to increase than to decrease, or more likely to decrease than increase, in response to the next random mutation. Thus, the response to the next random mutant in this special class of mutants, is on average, an oriented change in the pathways of differentiation, increasing flow down some pathways, inhibiting others. Under the drive of random mutants, the responding system tends, non-isotropically, in a given direction. Furthering the parallel with homeotic mutants, it is interesting that several of these in *Drosophila* transform eye to wing, none cause the reverse; several transform antenna to leg, none the reverse.

If our theory, and its systematic predictions about congruence of ontogenetic data (see below) are true, we must demonstrate some direct causal connection between the dynamics of ontogenetic stability and the dynamics of changes in ontogenetic programs. We believe the answer lies in what we call the principle of compensatory changes. We predict that any successful innovative change in any developmental program is accompanied by a countering, compensatory, change in all or part of the rest of the program.

Compensatory Changes and the Course of Evolution

If successful mutations are not random, but highly constrained, with respect to the entire genome, evolution would assume an intrinsic directionality. And if the emergence or fixation of each novelty was accompanied by increased variability in all or part of the rest of the genome, we would have a hierarchy-producing mechanism generating its own determinism *and* its own variation. The only other requirement would be abundant, nonlimiting energy and, because life originated and evolved within the energetic boundaries of this world, that would seem to be assured. Both Zuckerkandl (1976) and Kauffman (1973, 1983) appear to have partially anticipated this possibility. Zuckerkandl stated:

Any progressive directional evolutionary trend may be based on the progressive decrease in affinity constants between regulatory molecules and receptor genes. The decrease will be incremental especially in the case of mutations in receptor genes. Such a decrease may obviously be expected to be a universal mutational trend. This trend could be made use of in oriental (orthogenetic) evolution. *Over long evolutionary periods, it must necessarily be balanced, along most lines of descent, by corresponding increases in affinity constants, which occur perhaps elsewhere in the genome. Taken together, the decreases and increases probably account for most of the evolutionary transformation of organisms.* Yet the supposition can be made that the decreases correlate more generally with directional evolution, since they correspond to a spontaneous, thermodynamically oriented mutational trend. Repeated mutational "proposals," if they are not notably deterious, probably eventually become fixed in populations.

Kauffman (1973) drew similar conclusions about the significance of compensatory changes, or *coupled circuits* in his terminology:

If all subsystems underlying determination [in *Drosophila* development] were independent of each other the deductions made above [that some homeotic mutants may affect the stability of some circuits underlying determination] would hold. If circuits are coupled, a destabilized state of one circuit might be stabilized by couplings to other circuits in certain [imaginal] disks.

and summed up thusly (1983):

This result should not be surprising. Only ignorance in general would persuade us that the response of any integrated complex dynamical system to random alterations in its parameters or structure should be fully isotropic. *The tendency toward asymmetry may be common, and impose an internal force of unknown magnitude in evolution. Thus, orthogenetic tendencies may be common, if perhaps hard to recognize.*

This section will deal with evidence for the existence of compensatory changes occurring in systems of particular relevance to evolution and their expected effects.

We have suggested that ontogenetic programs comprise a relatively determinate part, which we term the *regulatory* or *canalized portion,* and a relatively stochastic part, which we call the *structural* or *noncanalized portion.* We may expect that mutations can occur in either portion, but that predicted effects of each type would be different. A successful mutation in the structural portion must integrate with the regulatory portion; likewise, a successful regulatory mutation must maintain the functional integrity of the rest of the regulatory network and the structural genes. Therefore, when looking for compensatory changes, one should examine the complementary portion(s) of the developmental program in which the mutation occurred.

For example, a successful structural gene mutation increases the variability of the structural gene portion; there is now an additional variable that may show up in descendent genotypes. If a single regulatory architecture is to account for all possible variants, such an increase in the structural portion will be compensated by a corresponding increase in regulatory determinism. This need not appear to be an active response by the regulatory architecture, but it will be manifested by a loss, in this case, of a degree of freedom associated with variability—it will be more difficult in the future for another mutant to be successful. It is likely that there is a threshold for the number of successful structural mutants that a given regulatory architecture can accommodate. Exceeding this threshold would result either in a breakdown of regulatory architecture, leading to aborted development, or to a compensatory reorganization of the regulatory architecture. If the entire lineage undergoes such a transformation, the result is anagenesis; if only a portion of the species experiences such changes, a portion of the ancestor speciates but the ancestor persists. As the accumulation of such mutants in a species approaches the threshold, the rate of change in the developmental programs of each species would be less and less, even though a high degree of heterozygosity might be demonstrated, say, electrophoretically, in each organism. And if the regulatory architecture includes even a single connection that acts as a very narrow channel through which a limited number of variants can pass during development, it is possible that a species' morphology may become "static" even though there has been no "temporal uniformity of the organism-environment relationship and/or genetic homeostasis" (Selander et al. 1970). Unable to break through the constraints of the regulatory architecture successfully, the species would persist, essentially unchanged, once a series of structural gene mutations had depleted the degrees of freedom permitted by the ancestral architecture, until extinction. Rather than selection eliminating all but a very narrow range of phenotypes or mutation rates decreasing for some reason, we propose that mutation rates would remain the same, but the rate of realization of new mutants would drop to near zero in time. We would predict a higher rate of aborted development in such species relative to their closest, "nonstatic" relatives. An excellent illustration of this point involves horseshoe crabs, "living fossil" arthropods. Selander et al. (1970) and Riska (1981) examined the amount of heterozygosity and polymorphism at allozyme loci and the degree of morphological variation, respectively, in *Limulus polyphemus*. In both cases, the variability found was in agreement with the general amount of biochemical and morphological variation found in other animals, both invertebrates and vertebrates. In addition, Sekiguchi and Sugita (1981) found that females of two Asian species of horseshoe crab, *Tachypleus gigas* and *Carcinoscorpius rotundicauda,* contained what they called "many bad eggs," which died soon after exposure to sperm.

The other aspect of the genetic basis of compensatory changes in evolution is that of mutations in the regulatory system. According to Zuckerkandl (1976), evolution might consist of stabilizing labile characters by fixing new concentration ratios or new relations in affinity constants between activators and repressors. Each such stabilization is expected to lead to some degree of labilization of *other* activator-repressor equilibria (creating an asymmetry; see discussions of Harrison in this chapter). Mutations result from decays in old affinity constants or increases in new ones. Decays in affinity constants realized in viable organisms usually would not proceed to a point of breaking integration patterns. They would be expected to modulate the steady-stage equilibria between interacting groups of controller genes. Decreases in affinity (= decreases in determinism or increases in stochastic part) figure prominently when shifts in regulation occur within a constant topology of the regulatory network. This agrees well with our conception of the relative ease, role, and compensatory effects of mutations in structural genes. However, Zuckerkandl also asserted that increases in affinity (= increases in determinism) between regulator molecules and receptor genes may be a prominent feature during buildup phases of genome organization, extensions, and recombinations of gene programs and of shifts in regulatory topology. Most such changes will still remain within the boundaries of stability (i.e., will not perturb the system to a new regulatory architecture), but some may be brought into the region of instability. A further slight change in the same direction will *then* produce some degree of reorganization of the portion of the regulatory network made unstable.

Despite technical difficulties in detecting and differentiating regulatory units, we may still identify classes of genotypic changes that exert regulatory influences and use them to test for evidence of compensatory changes. In general, we expect the evidence for compensatory changes resulting from changes in portions of the regulatory network to be more striking than that for structural gene mutations. Any change in the regulatory network will increase determinism in one part of the genome and also cause a measurable loosening, or increase in variation, in (an)other portion(s). For example, any mechanism decreasing the dissemination of genes from one chromosome to another in one part of the genome should result in increasing dissemination of genes in other parts. Current evolutionary theory does not provide for selection-induced increased determinism (= adaptation) for some traits, *coupled with* indeterminate increases in variation in other parts of the genome not affected by selection. If such phenomena exist, they would add a form of historical contingency (i.e., indeterminism) to a deterministic system.

Interestingly, data concerning this phenomenon have been known for over sixty-five years. Sturtevant (1919) discovered that when an inversion occurred in one chromosome of a homologous pair in *Drosophila* and crossing-over was halted or stabilized, the rate of crossing-over in the other chromosomes increased. These findings were considered anomalous in light of the principle of independent assortment. Interest in such *interchromosomal effects* was revived in the late 1960s (Lucchesi and Suzuki 1968). Studies on insects, primarily *Drosophila* and some orthopterans, and on corn (*Zea*) had demonstrated that interchromosomal effects were persistent findings in genetics studies. Inversions, translocations, and deletions of euchromatin material on one chromosome of a homologous pair stabilized or reduced the degree of crossing-over in that region but increased it in others (Suzuki 1973; Lucchesi 1976). Even deletion of some heterochromatin increases recombination rates in other chromosomes (Yamamoto 1979).

We would expect the same type of compensatory changes to occur even if only a minor rearrangement of the regulatory architecture was involved. Consider the case in which a new regulatory element is introduced into a genome. The chromosome on which it resides will be affected in part by the new element; the result will be a regulatory "stabilization" of the type envisioned by Zuckerkandl and by Kauffman. We predict an increase in variation in all or part of the rest of the genome. If the regulatory element is disseminated to another chromosome, say by translocation, we would predict a decrease in variability in the chromosome in which the regulatory element now resides, coupled with an increase in variability in the chromosome just vacated. The larger the number of regulatory connections, the smaller the likelihood of a successful shift in a regulatory element without causing a breakdown of the regulatory architecture. In terms of Kauffman's model, the more extensive the forcing structure, the less likely new mutations are to be successful and, at the same time, the more likely a successful mutant will produce a reorganization of the architecture.

Another aspect of compensatory changes enables us to understand how a species' potential information may be increased without significant alteration of the stored information system, that is, the ontogenetic backbone program common to all members of a species. Female *Drosophila* have two active sex chromosomes whereas male *Drosophila* have only one. And yet, structural gene activity on both autosomes and sex chromosomes is approximately the same for males and females. Such wholesale compensation occurs also in meta- and triploid females, intersexes, and meta-males (see Lucchesi 1977). This phenomenon is termed *dosage compensation*. Not all dosage compensations are viable, of course. Absence of one X chromosome in female horses results in Turner's syndrome. The "compensatory change" associated with this congenital problem is the failure to develop ovaries or a uterus. Clearly, this would not lead to the production of a new ontogenetic program. A similar phenomenon occurs at the genic level; it is then termed *gene duplication*. A gene may be doubled, and yet there need be no change in amount of gene activity for the doubled locus. One of the loci may not operate at all, or each may function partially, as with the sex chromosomes in *Drosophila*. Secondary loss of one of the genes would not alter the genome's structure; neither would dissemination of one of two genes to other parts of the genome. There have apparently been cases, especially among some plant species but also among animals, of *autopolyploidy,* in which the entire genome of homologous chromosomes is duplicated. The resultant offspring may be totally isolated reproductively from their diploid parents and yet exhibit very little divergence from them in terms of size, morphology, and activity of gene products (e.g., *Hyla chrysoscelis* [diploid] and *H. versicolor* [tetraploid] among the tree frogs of North America). It has been suggested previously (e.g., Kauffman 1983) that gene duplications may play important roles in evolution. This provides one answer to the question, "Where does new information come from?" (Wiley and Brooks 1982).

Compensatory changes similar to those discussed in the previous paragraph may also be seen in another class of reproductive outcomes, *allopolyploidy,* in which different genomes are combined and then doubled, thereby allowing viable offspring to be produced sexually. This occurs relatively often among plants, and in many cases, the offspring are virtually indistinguishable from one or the other parent; that is, they tend to segregate with one parent, showing little or no phenotypic evidence of being polyploids (see Funk 1981). This suggests compensatory inactivation of most or all of one parental genome, although there is no reason to assume loss of function of the functionally inactivated program. Thus, subsequent changes in the genome could uncover "new" variation that was already present. Because such "new" variation is derived from an ancestral coherent genome, it could not be viewed as undirected or independent.

Compensatory changes may also allow the emergence of new ontogenetic programs without changing the composition of the genome; the "text" does not change, but the "context" does. The addition of a regulatory connection (R) to one chromosome (A) alters expression in (an)other chromosome(s) as well. Subsequent translocation of the regulatory connection to another chromosome (B) returns the initial chromosome (A) to its original (prior to R) composition, but because the genome still includes the regulatory connection (R), the context of the expression of activity by A differs, and we would not expect a return to original expression. We find two different ontogenetic programs, one with R on A, the other with R on B, without requiring alteration of additional genetic material. And if the occurrence of R on A with subsequent translocation to B results from a thermodynamic symmetry breaking, the change in "context" would be enough to ensure that we would not expect to find the opposite transitions to occur (i.e., R on B $\longrightarrow$ R on A). There is growing evidence that large chromosome fragments may be disseminated (see Bush 1975b; Baker and Bickham 1980). The result is a clade containing many differentiated species, each karyotypically distinct, with each karyotype being a "shuffled" version of a single ancestral karyotype. Baker and Bickham (1980) referred to "extensive *and* conservative" [emphasis ours] karyotypic changes in the evolution of karyotypes of various chiropterans (bats). The fragmentation of chromosomes is compensated by a consolidation of the pieces into new chromosomes. Of course, not all chromosome fragmentations result in viable new programs. There are a number of medical and veterinary syndromes characterized karyotypically by fragmented chromosomes or fragmented chromosomes with abnormal reattachments. Thus, to reiterate an earlier point, despite the fact that many genic and chromosomal reshufflings can occur, the only ones which will be realized are those which can be integrated into the rest of the genome without interfering with viable development. If the entropically favored mutations are those which are incompatible with at least part of the developmental program for a given species, the result will be a "static" species with unusually high rates of aborted development.

An additional mechanistic basis for genetic compensatory changes has been delineated, at least partially. Two factors are involved. The first is the randomizing, or stochastic, effects of mutation, gametogenesis (meiosis), and sexual reproduction. Coupled with this is the deterministic trend toward genome homogenization called *molecular drive* (Dover 1982 and references therein). The genomes of all eukaryotic organisms examined thus far include a large complement of redundant genetic sequences, the so-called multigene families. These tend to be highly homogeneous within species and discrete among species. In addition, many of the multigene families have "subfamilies" on different chromosomes, including nonhomologous chromosomes. Three documented genetic mechanisms can produce such configurations. These are (1) unequal exchange of genes across chromosomes, (2) gene transpositions and duplications, and (3) biased gene conversions, either on the same chromosome, across homologous chromosomes, or even across nonhomologous chromosomes. Molecular drive refers in general to all such slow-acting deterministic factors affecting genome turnover. Koch (1983) also suggested that increasing numbers of copies of genes should slow evolutionary rates. The interaction of molecular drive with the relatively fast, stochastic factors of single-copy gene mutation, meiosis, and sexual reproduction should produce a coherent increase in genetic complexity locally and globally through time.

Because the multigene families have representatives on different, nonhomologous, chromosomes we have additional evidence that the genome acts as a single unit. As molecular drive acts to homogenize portions of the genome across nonhomologous chromosomes, we would expect to see an increase in stochastic fluxes in other parts of the genome. This provides a mechanistic connection between the observation of interchromosomal effects and the principle of compensatory changes.

Up to this point, we have discussed compensatory changes in genetic terms only. Cytoplasmic factors may play a significant role in development as well (Løvtrup 1974). In our view, this is not unexpected. In multicellular animals, at least, the cytoplasm of any fertilized egg is the cytoplasm of the oocyte, and it is highly organized before fertilization. Any new genetic combination must integrate with this initial cytoplasmic organization. This provides an additional set of intrinsic constraints on the realization of viable descendent organisms. In terms of compensatory changes, we would expect the interaction of cytoplasm and genome to result either in a reduced range of expressed gene products at a particular time relative to the possibilities of the genetic architecture then existing, or we would expect cytoplasmic changes to occur concomitantly with genetic differentiation. A "mutual compensatory-change" system between cytoplasm and nucleus would provide a feedback system leading to great stability. An example of this sort of phenomenon is the syndrome known as *hybrid dysgenesis*. When males of certain laboratory strains of *Drosophila* are mated with certain wild-type females, normal F_1 progeny result. However, when females of the laboratory strains are mated with wild-type males, some of the F_1 progeny show a variety of "dysgenic" anomalies, including male sterility and reduced fecundity, increased recombination rates, and male recombination. The wild-type genetic complement contains elements called *transposons*, which under certain conditions are capable of moving from one chromosomal position to another, producing the dysgenic conditions. The female cytoplasm of some laboratory strains provides such conditions, although it is not definitely known whether the cytoplasm contains transposon inducers or lacks transposon inhibitors. It is known that transposons do not move about randomly but tend to move to particular chromosomal regions, which are termed "hot spots." Some laboratory studies have shown that transposons tend to replicate as they move about, increasing the amount of gene product transmitted to the cytoplasm. When the number of transposon copies reaches thirty to fifty per genome, the cytotype changes and the dysgenic syndrome disappears from laboratory populations. It appears that a threshold level of gene product, produced by thirty to fifty copies of a transposon, can alter the cytotype, producing a self-stabilized steady state. For a review, see Engels (1983) and Kidwell (1983). In fact, for "reactions" that occur in the nucleus and "diffusion" that occurs in the cytoplasm, this could be the cellular context for Harrison's "short-range activation/long-range inhibition" model. We might even postulate a major role for cytoplasmic effects in explaining why most allopolyploids in plants resemble one parent rather than appearing as a mixture of both.

The principle of compensatory changes may allow us to provide new explanations for some macroscopic phenomena. For example, if evolutionary change produces increasing amounts of determinism in any evolving system, why do we find essentially equal amounts of genetic polymorphism and heterozygosity at structural gene loci among members of a clade, or even across clades (e.g., horseshoe crabs and vertebrates)?

For biological evolution, a constant or decreasing rate of successful mutation would produce descendent lineages in which there was progressively less variation. We do not see this; even in "evolutionary dead end" species such as horseshoe crabs, there is a significant amount of variation, at least in structural gene products. Therefore, the explanation we seek must involve an increase in rates of successful mutation, either continual or periodic. If continual, we must find a mechanism that postulates increasing rates of successful mutations across *all* lineages of organisms regardless of the nature of their particular developmental programs. In light of evidence presented earlier in this chapter, it seems unlikely that this occurs. If the increases are episodic, we must provide a mechanism accounting for periodic increases in variation in each lineage under a specified set of conditions. We propose that set of conditions is any episode of increasing determinism, that is, any change in the regulatory architecture of a species. In this manner, the evolutionary process is one that remains organized because it is becoming progressively determinate historically, but remains open and indeterminate with regard to future occurrences because the emergence of new determinism also causes a change in boundary conditions, allowing new variation. Zuckerkandl (1976) seemed to have a similar view of the evolutionary process:

> a stabilization of one morphological character can and, on our assumption, will bring about a destabilization of some other morphological character, because in terms of this other character the equilibrium between some regulators and receptors will now be somewhat closer to the flip-over point of this character, at the brink of this flip-over point, or beyond this brink.

Zuckerkandl suggested that cases of terminal deletion and substitution occurred commonly in transformations from free-living to sessile life-styles, including parasitism. He suggested that the rhizopod crustacean *Sacculina,* a parasite of crabs, was a good example. The larval stages of *Sacculina* are typical free-swimming nauplius and cypris forms similar to their free-living relatives. However, the adult form is essentially an amorphous saccate structure containing genitalia and exhibiting rootlike extensions growing into the visceral mass of the crab. Zuckerkandl suggested that loss of terminal portions of the ontogenetic program, that is, the ones associated with free-living functions, would not adversely affect the life of a parasite and that the saclike body of the parasitic adult represents the emergence of a compensatory regulatory structure encompassing structural genes previously associated with formation of other structures. A similar kind of phenomenon occurs during the life cycle of the copepod *Haemocera danoe*. The nauplius stage is free swimming, but the next larval stages are not present; the nauplius penetrates a serpulid worm and becomes a small mass of nondifferentiated cells. This mass then develops into the adult form, which is a typical free-swimming copepod. Loss of nonterminal information results in a nondifferentiated mass, but in a compensatory manner, new regulatory architecture emerges and the adult is produced. It is pertinent to note that the cypris larva of *Sacculina* contains a mass of undifferentiated cells that are injected into the crab host and become the adult form.

There are a number of manifestations of the principle of compensatory changes that emerge in conjunction with specific organism-environment interactions. We have seen previously that dissipative structures in general are characterized by alternate dissipative pathways connected to each other by feedback loops. Such systems are by nature compensatory. Those which we will now consider belong to the class of phenomena Zotin and Zotina (1978) termed *inducible* processes, both *pulsed* and *adaptive* (see pages 78–79). It is important that they be recognized as compensatory changes because they might otherwise be misconstrued as examples of environmental selection.

Energy taken up by a living organism from the surroundings may be utilized for one or more of four general processes: (1) growth, or increase in size complexity; (2) differentiation, or increase in heterogeneous complexity (Wicken 1979); (3) homeostasis, or maintenance of self generated boundary conditions; and (4) reproduction, or increase in population complexity of the species. The principle of compensatory changes requires an increase in activity in one or more of these pathways if another is blocked.

For example, the termination of growth and differentiation is often highly correlated with the initiation of reproductive activity. Likewise, organisms that are small, with relatively short developmental periods, have high metabolic and reproductive rates, whereas large organisms, with long developmental periods, have relatively low metabolic and reproductive rates. For example, Pechenik and Lima (1984) found a compensatory relationship between growth and differentiation and larval life span in a species of gastropod mollusk. Those larvae which grew more slowly remained larvae longer than those which grew more quickly, although there was no evidence that this affected adult life span or reproductive output. Levin (1984) demonstrated two reproductive modes in a polychaete species. In the first mode, females produced few large eggs, which developed into larvae that settled to the bottom quickly. In the second mode, females produced many small eggs, which developed into larvae that became planktonic for a period of time before settling. O'Grady (1982) noted that this second example represents an example of a universal truism under our theory, but is accorded selective significance under current evolutionary theory, as the principle of "K and r selection." It is true that the first reproductive mode produces K-selected larvae, which settle early and do not disperse widely and are relatively larger than free-swimming forms, the attributes of any species that will survive intrusion by a colonizing competitor. It is also true that the second reproductive mode produces r-selected larvae, which disperse widely and occur in large numbers, displaying the attributes of successful colonizers. These ecological capabilities, as important as they are to the biology of the species, are effects, or emergent properties, of compensatory developmental changes and are not responsible for the developmental patterns. Function follows form and not the reverse.

We may see physiological manifestations of the principle of compensatory changes. Immediately following amputation of the tail, axolotls show a short-term intensification of respiration and glycolysis. Zotin (1972) reported that maximum glycolysis occurs within two hours and maximum respiration within sixteen hours. By twenty-two hours after amputation, rates of glycolysis and respiration have dropped to their preamputation levels and, in a compensatory manner, regeneration of the tail structure has begun.

Harrison (1981) predicted that morphogenetic pathways driven by chemical kinetics should be temperature sensitive. Perhaps the most striking evidence for this type of compensatory change comes from the study of sex determination in amniote vertebrates. Among these animals, temperature-dependent sex determination has been documented for at least some turtles, crocodilians, snakes, and lizards.

In fact, it is possible that only those amniotes with structurally differentiated (heteromorphic) sex chromosomes do *not* exhibit such a mode of sex determination. Harvey and Slatkin (1982) recognized that temperature-dependent sex selection would have a definite impact on the population biology of any species having the trait. However, they sought the explanation for why this trait evolved in a discussion of the effect on population biology the trait exerts, rather than looking for the mechanical causal agent that allows such a trait to exist in the first place.

Another set of observations that has seemed to require an environmentally mediated selectionist explanation is the evolutionary trend, most apparent in salamanders, called *neoteny*. For all known amphibians, metamorphosis from the larval, or "tadpole," stages to the adult is stimulated by thyroxin produced by the larva's thyroid gland. Removal of a larval frog's thyroid results in a very large tadpole. Removal of the thyroid inhibits continued differentiation, but the organism's developmental program compensates and the result is the larger tadpole. Alternately, giving extra amounts of thyroxin to a normal anuran tadpole stimulates early onset of metamorphosis; the resulting metamorphosed frogs are much smaller than normal (Weichert 1970).

Development of some caudate amphibians (salamanders) differs from that of anurans in an important way: the genetic program for sexual maturity and that for metamorphosis to the adult body form are independent. In anurans, removal of the thyroid rudiment from larvae produces a large larval form that never becomes sexually mature. In salamanders, this is not always true. In five separate families of caudate amphibians, sexual maturity occurs in larval forms, and metamorphosis does not regularly occur. For some of these, metamorphosis may begin but not be completed, and in others it is not even initiated. For a certain group, thyroxin injections may initiate metamorphosis, but for others even high doses of thyroxin will not produce any changes. In all cases, thyroxin is produced endogenously, and thyroid extract injected into other salamanders will stimulate metamorphosis. Thyroxin apparently combines with an activator, which then undergoes an allosteric transition enabling it to combine with a receptor gene. This represents a form of "flip-over" in Zuckerkandl's terminology. In at least five different cases, an apparent mutational decay in some affinity constant within the controller system has occurred. The effect of this is to raise the threshold of sensitivity to thyroxin. In some cases, such as *Amphiuma, Siren,* and *Necturus* (Amphiumidae, Sirenidae, and Proteidae), the mutational decay has been severe enough that the adult gene program has subsequently decayed and metamorphosis can no longer be elicited. Like thyroidectomized anuran tadpoles, these salamanders are very large relative to regularly metamorphosing forms.

For some of these neotenic salamanders, metamorphosis occasionally occurs in natural populations. Their gene program for metamorphosis is still intact and their mutational decay in thyroxin sensitivity has been mild. Members of such species which, for genetic or environmental reasons, possess very high levels of thyroxin will metamorphose. The best studied cases of this sort of situation are those involving ambystomatid salamanders such as *Ambystoma mexicanum,* whose neotenic forms are known as axolotls (for a review of their development, see Abeloos 1956). Because the sexual maturation program is independent of metamorphosis, both the axolotl form and the metamorphosed forms achieve sexual maturity in the same amount of time after hatching. Therefore, if the principle of compensatory changes holds for the four alternate dissipative pathways listed previously, any truncation of differentiation, as shown by axolotls, should be accompanied by a compensatory increase in the growth or metabolic rate of the organisms exhibiting the truncated development. Axolotls routinely achieve sizes greater than twice that of their metamorphosed siblings. Among salamanders, increased body size associated with neoteny, or ontogenetic truncation, is clearly not an adaptive response in the Darwinian sense. It is ultimately a deterministic compensatory outcome of the decay of an affinity constant for thyroxin, which may proximally have adaptive consequences.

One implication of our discussion concerns the methodology of discovering episodes of neotenic speciation. It has been argued that if a species' ontogenetic program has been formed by terminal addition, that is, recapitulation, neotenic speciation will produce new species that are identical structurally to remote ancestors. This might be true if it were known that (1) developmental programs had evolved by means of recapitulation *and* if it were known beforehand that (2) neoteny had occurred *and* if it were further known that (3) the principle of compensatory changes was falsified. So long as compensatory changes are a universal aspect of all evolutionary change, there will be unambiguous markers of phylogenetic change even if neotenic speciation and simple recapitulation occurred. Thus, we might expect to find neotenic *parts* of species but not neotenic *species.* However, we contend that all three of these criteria can be assessed only a posteriori following a phylogenetic analysis (Fink

1982; see also chapter 5); thus, one need not be concerned that neotenic speciation, if it occurs, will require special techniques to detect. Of course, if one has empirical evidence stating that neoteny is not feasible due to the *inter*dependence of sexual maturation and terminal differentiation, one would not wish to postulate neotenic speciation at all.

Within-population polymorphisms in the rotifer *Asplanchna sieboldi* have been explained in strictly selectionist terms (see Gilbert 1980 for a review). What is observed is this: *A. sieboldi* exhibits three different morphs, which we will call I, II, and III. Changes in dietary conditions can evoke a transition from I to II or from II to III. Morph I is produced by asexual reproduction of other I's or sexual reproduction of morphs II and III; it is the smallest morph, is saccate in shape, does not reproduce sexually, and exhibits the highest rate of asexual reproduction of the morphs. Morph II is produced by morph I and itself; it is intermediate in size, cruciform in shape, and intermediate in position in the developmental sequence. It shows the highest degree of sexual reproduction and lowest degree of asexual reproduction. Morph III is produced by II; it is the largest, is bell shaped, and exhibits intermediate reproductive capabilities. Although somewhat more complex than the previous examples, this clearly shows the compensatory nature of *A. sieboldi*'s polymorphism. Every organism in the species carries the genetic potential for all three morphs but, depending on the morph of the parent and the dietary conditions, only one morph is expressed. The particular expression is an inducible process (*sensu* Zotin), and even though the inducer is extrinsic to the organism, the morph that is induced is not an evolutionary adaptation to the inducer but a compensatory modification of the previous morph—no genetic change has occurred. Each morph is a differential expression of the same potential information.

Historical Analysis and Ontogeny

Systematic documentation of predicted classes of ontogenetic transformations during evolution is a necessary first step in testing developmental predictions of our theory. Systematic analysis can provide four sets of observations that are expected to occur if our theory has validity.

First, we should find that the most efficient summary of characters will be hierarchical and not a periodic table. We do not mean to insult our reader's intelligence by this statement, but it is all too often forgotten that the presumed occurrence of a natural hierarchy of form among living organisms prompted evolutionary speculations in the first place.

Second, it should be generally true that phylogenetic trees obtained using characters from different developmental stages will still predict the same set of phylogenetic relationships for the included species. Furthermore, when absolute congruence is not found, the relationships postulated by each set of characters will all be *consistent* (*sensu* Wiley 1981b) with each other. This would mean that evolutionary changes in ontogenetic programs were better correlated with the history of the evolving lineage than with particular adaptive/selective regimes to which organisms at different developmental stages are subjected. This is expected because it is intrinsic self-organizing phenomena that *effect* ontogeny; extrinsic selective factors may only *affect* ontogeny. Adaptive changes in developmental programs will comprise only those which do not conflict with the historical, self-organizing core. A few such studies have been done (e.g., Hennig 1950b, 1953, 1969; Howden 1982; Pinto 1984; Brooks, O'Grady, and Glen 1985), and they support this prediction.

Third, we should be able to find cases in which the evolutionary change in a developmental program did not involve terminal addition (i.e., recapitulation). For example, parasitic flukes, which are flatworms called digeneans, have life cycles characterized by a number of larval stages unique to them, and yet the adult digenean resembles many other kinds of flatworms. Clearly, the larval stages have arisen after the origin of the adult form and have been intercalated into the developmental program. The relative stability of the adult stages despite such nonterminal developmental changes indicates that the whole developmental program is a coherent system. If ontogeny is such a causal sequence, evolutionary changes may occur at any point in the sequence, but they will always be modifications of the preceding steps (temporal constraints), and only those compatible with the rest of the sequence will survive (developmental constraints). This unity, or cohesion, of the developmental program provides a set of intrinsic constraints on the realization of evolutionary novelties. And if all developmental stages predict the same phylogenetic tree (the second point cited), we can deduce that the constraints of the unity of the developmental program are themselves products of and constrained by past history as well as immediately preceding developmental programs.

Finally, we think that the process by which new structures emerge from old in development is analogous to that by which new species emerge from old. If that is true, then the technique for systematic analysis deduced from our theory (chapter 5) should also be useful for mapping the fates of cell lineages in ontogeny. There are hints in the literature (e.g., see Katz and Goffman 1981) that ontogenesis might involve efficient structural transitions. However, there are currently no published studies examining this question explicitly.

Summary

In this chapter we have attempted to show that all four principles of our theory, *irreversibility, individuality, intrinsic constraints,* and *compensatory changes,* characterize biological development. We have presented evidence that ontogenesis is a nonequilibrium phenomenon, that organisms are dissipative structures whose temporal sequence of change is caused by an interaction between relatively determinate (regulatory) components and relatively stochastic (structural) components within the genetic and epigenetic architecture each organism possesses. The genome acts as a coherent unit, thus constraining the range of possible mutations to those which can integrate with the entire ontogenetic system. We have further provided examples of types of compensatory changes suggesting that new variation in species is the result of episodes of increasing determinism. Diversification of ontogenetic programs occurs as a result of entropic change promulgated by reproduction. In the next chapters, we will see that a further amplification of this phenomenon results in the production of new species. Thus, the emergence of new species follows an analogous process as the emergence of new structures in ontogenesis.

Because changes in developmental programs are constrained by the parental regulatory architecture, some mutations will be entropically favored and others will be prohibited. Evolution of developmental programs will always be directional. However, this directionality will not be teleological, as assumed in the past by opponents, and some proponents, of orthogenesis. There is a difference between "directional" and "deterministic." A certain directionality is imparted by the fact that change is made in reference to history. Evolution is indeterminate as far as the future is concerned, even though evolutionary trajectories that we see today have been determined by past trajectories. With each increase in deterministic factors, new variation is generated. In this manner, biological evolution is a self-organizing and indeterminate process.

This establishes a new context for the notion of "adaptation" in biology. If the system is at an equilibrium, or steady-state, point, it will react to extrinsic stimuli by tending to remain at equilibrium. If it is some distance from equilibrium, it will react by tending to reduce its rate of entropy production. For living systems, responses to changing conditions will be directed along these entropic lines first. We should thus expect to find a class of evolutionary innovations that (1) characterize highly successful lineages and (2) have the effect of minimizing entropy production. One such class of innovations is the trend toward increasing complexity and self-regulation in developmental programs in all lineages of living things. Others should be sought in long-term evolutionary trends involving ontogeny, including the subject of "static" species.

We have also provided testing protocols and research programs for those interested in pursuing these views further, on an empirical basis. The protocols often involve a combination of systematic analysis and developmental biology, an alliance in biology previously reserved for recapitulationists. By showing that ontogenesis is a causal sequence, we have predicted systematic congruence of characters from all stages of development, even if evolutionary changes have not involved recapitulation. We have attempted to pinpoint relevant areas of research for those interested in formulating a unified theory of development. Although we concentrated primarily on the work of a few authors, we recognize that there are many other authors who share part or all of the world view we present. Our intention has been to pinpoint threads of continuity among different conceptual approaches to the study of developmental biology.

The viewpoint developed in this chapter may be expressed as a series of predictions, for some of which no testing protocol yet exists:

1. Every evolutionary change allows a new but finite class of mutations. Evolutionary progress is directional (orthogenetic) but not determinate or teleological. This emergent self-organization will be the predominant aspect of evolution; the amount of extrinsic (i.e., selective) pressure required to effect a change in these intrinsically generated trajectories will be massive, and seldom, if ever, realized.

2. All successful ontogenetic "deletions" will be accompanied by a compensatory "substitution"; mutations affecting morphogenesis produce coordinated modifications of inductive effects because new states must be integrated with preceding states of the system.

3. Compensatory changes are the real essence of epigenesis, and compensatory changes have an irreducible historical element. Non-compensatory changes are nonregulatory and hence evolutionarily neutral.

4. Widespread congruence of data from all ontogenetic levels is expected in systematic analyses regardless of the mechanism or point of ontogenetic change in each case. This is because ontogenesis and changes in ontogenesis are causal sequences that change through time but maintain continuity through reproduction.

5. To the extent that changes in developmental programs indicate evolutionary changes in lineages, we would expect to see the rather sudden (in geological terms) appearance of an epiphenotype, corresponding to a new regulatory architecture, followed by a relatively long period in which relatively minor changes occurred in structural components within the regulatory architecture, to the extent that the regulatory architecture permits. Emergence of a new regulatory architecture will be accompanied by a corresponding, rather sudden, disappearance of the previous regulatory architecture.

· 4 ·

Populations and Species

Evolution is not just a manifestation of changes in the genes and developmental programs of individual organisms. It is also a manifestation of the way in which such changes, reflected in phenotypes, affect populations and species. In this chapter, we move from the level of the individual organism and its ontogeny to the levels represented by populations, species, and groups of species. We wish to show that mechanisms at work in populations are entropic processes and that our reasoning can be extended to species and speciation.

Evolution in Populations

Within-species interactions represent an interface between processes at the level of individual organisms and the process of speciation. Evolution within species is the domain of the discipline of population biology. Important fields of study included in population biology are population dynamics, population genetics, autecology, and synecology. We are not population biologists. Fortunately, much of the vast body of literature in this field has been summarized in a number of readable works, including Wright (1968, 1969, 1977, 1978), Crow and Kimura (1970), Dobzhansky (1970), Wilson and Bossert (1971), Kimura and Ohta (1971), Lewontin (1974), Endler (1977), and Futuyma (1979).

Thermodynamic Considerations

It is relatively easy to show that individual organisms are dissipative structures (chapter 3). Might populations also be dissipative structures? If so, how does their dissipation differ from simply summing the dissipation of the members of the population? Populations can hardly be considered dissipative systems unless we can identify some dissipative function or functions associated with populations that are not associated with single individuals. We suggest that at least some populations can be considered dissipative structures because they comprise reproductive communities and thus can transform energy, in the form of offspring, which remains in the system longer than the life span of any individual.

Let us consider a population that maintains a constant number of individuals year after year. The average flow of energy into the population is constant and reflects the ability of each member of the population to acquire and transform energy, either in the form of photons or in the form of food. The death of an individual represents a loss of energy-processing capability by the population and a decrease in the flow of energy from the environment into the population. The birth of an individual represents a potential gain in the ability of the population to process energy. If births equal deaths, as in this example, the entropy production of the population will be relatively constant, representing the average metabolic activity of the population. Thus, there will be a trade-off between energy acquisition and the sum of energy used to do work, energy lost as heat or excreted, and energy converted to structure. Such a population occupies a steady state of minimum entropy production. It maintains itself by reproduction.

What happens if the number of individuals composing the population increases or decreases? The population can increase only if energy is available beyond that already being used. If additional energy is converted into more individuals who grow to maturity, there will be an increase in entropy produced within the population. Decreases in the number of individuals will have the opposite effect. We may express this relationship by considering changes in the thermodynamic entropy of the population (dS), as shown in fig. 4.1. Note that the total entropy production within the population parallels the increases and decreases in the number of individuals. Although population decline results in a decrease in the *rate* of entropy production (amount of entropy produced per unit time) within the population, d_iS is still positive because all organisms must carry out metabolic activities, and unless reproduction is stopped completely, entropy is produced as a result of energy invested in reproduction and growth. So long as there is a single organism alive in a population, there will be some level of entropic behavior. And, so long as some reproduction is occurring, there will be entropic behavior

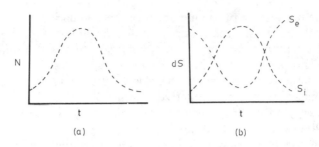

Fig. 4.1.   The relationship between numbers of individuals over time (*a*) and rate of thermodynamic entropy production over time (*b*). S_e = Entropy production resulting from energy flows from the environment; S_i = entropy production resulting from changing numbers of organisms in the population.

at a higher functional level than that of individual organisms. Also note that there is an opposite increase and decrease in the overall rate of entropy flow from the surroundings, which reflects the availability of matter and energy to entropy-producing systems outside the population we are considering.

Although it is possible to view populations and species as thermodynamic entropy systems, energy is not a determinate capacity and will not provide us with the kinds of insights into the evolutionary processes that we wish to discover. The deterministic capacities at this functional level are information and cohesion.

Information

At the level of populations it is not so important to know how long and complicated any one organism's ontogenetic program is or how many structural gene loci are present. Rather, it is important that variation exists among individuals, because variation is necessary for evolution to occur. The more variation in the heritable information of a population, the more complex that population is. The more variation among populations, the more complex the species is.

The ultimate sources of heritable variation are allelic and chromosomal mutations. The effects that mutations have on the phenotype range from nil to what can be described as saltatory. We consider the occurrence of such changes to represent an increase in the number of microstates accessible to the information system of the population and, by extension, to the species as a whole. According to the evolutionary dynamic discussed in chapter 2, an increase in the number of genetic microstates in a population should cause an increase in the entropy of the population's information. Let us consider an increase in the number of microstates to be indicated by an increase in A, the number of microstates accessible to the population. When A increases, the entropy of information, or the complexity of the information, increases.

The maximum evolutionary potential of a population is log A, representing the maximum number of possible microstates. This is also the maximum entropy possible given A. For example, if the population is a "monotone" homozygous for all loci, there is one accessible microstate ($A = 1$) and no possible production of information entropy, since $\log_2 1 = 0$. To see how information behaves in the simplest case, consider a single locus in a population composed of fifteen individuals. Beginning with all individuals being homozygous for the same allele x, the frequency of a new mutant x' gradually increases. The origin of x' increases the entropy of the population from minimum (in this case $H = 0$) to

$$H = -(29/30 \log_2 29/30) - (1/30 \log_2 1/30)$$
$$= 0.210 \text{ bit}$$

where 29/30 is the probability of finding the allele x in any member of the population and 1/30 in the probability of finding x' in any member of the population. The maximum value of H given the new mutant can be calculated by finding log A. For any mutational change, there are initially two genotypes (xx, xx') and maximally three genotypes (xx, xx', $x'x'$). We shall assume for simplicity that no additional kinds of mutations occur, that is, that no x'' or x''' derives from x or x'. Thus, the maximum complexity as a function of statistical entropy would be

$$\log A = \log_2 3$$
$$= 1.585 \text{ bits}$$

This value would be achieved only if at some point in the evolution of the population there was an equal distribution of x and x' among members of the population such that the probability of finding either was the same. In standard notation, this would occur when $p = q = 0.5$, where $p = p(x)$ and $q = p(x')$. There are two configurations of genotypes in which this can occur: (1) if an equal number of individuals is distributed among the three genotypes or (2) if an equal number of individuals is distributed among only two genotypes, xx

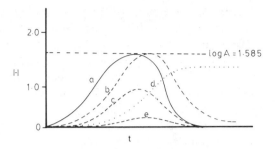

Fig. 4.2. Changes in the statistical entropy of a population for two alleles over time, assuming that the population begins as a homozygous population for one allele, x, with x' originating after the time represented by the intersection of the x and y axes. The shapes of the curves are entirely heuristic. H = Entropy; t = time.

and $x'x'$. If we imagine a population drifting toward fixation for x', it is possible to reach the condition described in (1), which is the maximum complexity possible for the population. If we represent this complexity in terms of information entropy, we find that

$$H = -(10/30 \log_2 10/30) - (10/30 \log_2 10/30) - (10/30 \log_2 10/30)$$
$$= 1.585 \text{ bits}$$

In this case, $H = \log A$, the maximum entropy possible given the three genotypes. Of course, other things could happen. Subsequent drift might result in fixation for x or x'. If so, the population will show a decrease in H, reflecting the disappearance of one allele. Complete replacement of either x or x' will result in the population's returning to

$$H = -(30/30 \log_2 30/30)$$
$$= 0$$

And, once the alternative allele disappears, A will return to $A = 1$ and $\log A$ will also be 0. This would indicate no net increase in information entropy. There are still other outcomes. Suppose that either selection or a biased mutation rate for x' occurs. It is possible that the population would never reach maximum complexity before H values began to decrease. And if a polymorphism is established, the H value will never return to 0. Some of the kinds of change we should expect with such models are shown in fig. 4.2. In fig. 4.2a, we have shown a drift model where the species actually reaches the value $H = \log A$ and no back mutation to x occurs. Fig. 4.2b represents a no-selection model with recurrent mutation such that an equilibrium is reached; the value of the equilibrium can be calculated by the following:

$$f(x) = \frac{u}{u + v}$$

where $f(x)$ = the equilibrium value of the allele x and u and v are the forward and back mutation rates, respectively. The H values would be calculated using x, in conjunction with information on the distribution of mutant genes among individuals. Fig. 4.2c might represent selection against x or a nonselection model based on biased mutations favoring x'; that is, the mutation rate of $x \longrightarrow x'$ is faster than the rate of drift to fixation of x', and back mutations occur at a much lower frequency than the forward mutations. Fig. 4.2d represents a balanced polymorphism. Fig. 4.2e represents selection against x' but assumes recurrent mutations so that the equilibrium point is established in a manner similar to fig. 4.2b, but using the mutant gene as the numerator.

During any such equilibrium phase, the entropy of the population can fluctuate between the ancestral condition and a maximal state determined by log A. Note, however, that there is never a decrease in entropy below the ancestral level. Trivially, this means only that evolution does not destroy history, that is, it does not run backward in time. At any point in time, with reference to its ancestor, a population's information entropy is greater than or equal to zero. Because historical entropy cannot be destroyed, no negentropic behavior occurs. And, in accordance with Hollinger and Zenzen's (1982) characterization of dynamic entropy production in an equilibrium state, we find nothing inherently irreversible in the dynamics of an equilibrium population. Thus, no net entropy production is expected so long as the population remains in an equilibrium state; any apparent entropy changes are reversible fluctuations around the equilibrium point. However, irreversible changes that might occur would be marked by an increase in entropy and the emergence of macroscopic organization.

We do not expect macroscopic organization to emerge spontaneously from an equilibrium state. Rather, such organization is expected to emerge whenever the entire system is in a dynamic nonequilibrium state. Two types of forces can contribute to such a phenomenon: those extrinsic to the system that are strong enough to force an irreversible change and those intrinsic to the dynamics of the system that are themselves inherently irreversible. The in-

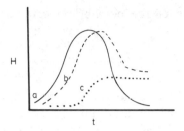

Fig. 4.3.   Heuristic entropy curves for three different types of anagenetic change.
a = Descendant populations that become highly complex, then secondarily return
to complexity no greater than starting point; *b* = descendant populations that become
highly complex, then less complex, but remain more complex than starting point; *c*
= descendant populations that become more complex, stabilizing at highest complexity.

formation presented in chapter 3 leads us to believe that there are
inherently irreversible processes affecting the information content
of populations. These processes include aspects of reproduction and
mutation, especially aspects of concerted evolution and nonrandom
production of mutants, as well as aspects of selection.

Consider two variable loci or a locus with more than one allele.
The number of accessible microstates increases and the chance of
a species reaching log A decreases. For the one-locus/three-allele
model, A = 6 and log A = 2.585 bits. For the two-locus/two-allele
model, A = 9 and log A = 3.170 bits. For the species to be maximally
complex, the alleles in either case would have to be distributed in
an equiprobable manner, with the same number of individuals for
each genotype. This is unlikely, even with drift, since both loci or
all three alleles would have to drift the same way at the same rate.
When we consider all variable alleles, we can gain an understanding
of why evolution is a constrained process and the true log A value
(i.e., equilibrium) is never approached.

Some mutants that arise in a population may increase the complexity of the part of the ontogenetic program in which they occur.
What is the shape of the curves generated when there are entropic
changes in the amount of information needed to specify a mutant
phenotype? We (Wiley and Brooks 1982) presented three general
curves for anagenetic change, which are summarized in fig. 4.3. Fig.
4.3 represents the kinds of curves we would expect for a measure
of information change in the context of population biology. In our
first presentation (Wiley and Brooks 1982), we considered these
curves to represent pure information changes, but we were in error.
Because it is impossible for a mutation to spread instantaneously,

the rate of change, represented by the slopes of the curves, embodies a cohesion parameter. We will consider this in the next section. The peaks in fig. 4.3a,b represent the maximum complexity that the population achieves, represented by equal occurrence of all phenotypes. The decrease in rate of entropy production on the backside of the curve represents the reduction in complexity as one of the genotypes or phenotypes drops in frequency of occurrence. The intrinsic increase in complexity of a character is represented in 4.3b by the fact that the curve does not descend to the ancestral entropy level but stabilizes at a higher entropy level that reflects the larger amount of information needed to specify the new phenotype (more information is dissipated during ontogeny). Fig. 4.3c represents the polymorphic condition; this particular curve does not necessarily imply additional increases in intrinsic complexity of the novel trait.

Overall Estimates of Information Entropy

When we consider the entire genome of each member of a population, measuring the population's complexity in terms of statistical entropy must be complicated indeed. For example, a ten-locus system with two alleles per locus gives 59,049 (3^{10}) possible genotypes, and we would have to census the population to determine its entropy. While such a thing might be conceivable for ten loci, estimating an overall H for any population of organisms having more loci would seem impractical. However, it has been possible to pursue this question through the use of computer-assisted simulations.

At this point we will discuss a quantitative model to demonstrate the connection between entropy increase and information increase in biological systems first introduced by Brooks, LeBlond, and Cumming (1984).

The model incorporates both cohesion and information components in terms of the rules governing the time evolution of strings $x_i(t)$, $i = 1,2,\ldots,N$, whose elements represent the stage attained by a sequence of discrete developmental stages (the model uses $N = 9$). Each element represents, minimally, two homologous alleles. Each string represents an epiphenotype, or developmental program (i.e., an individual organism). A species is thus a population of such strings. At each time step, there exists a finite probability, $P_m(i)$, that a mutation will occur at the ith stage of ontogeny. For simplicity, it assumes that the mutations produce instantaneous replacement without going through a polymorphic stage and that the earlier in the ontogenetic program a mutation occurs, the less likely it is to be successful. Hence, $P_m(i)$ is very small for $i = 1$ and increases with i, reaching a maximum at $i = N$. The program uses

$$P_m(i) = 2^{-N(1 + 1 - i)}$$

so the probability of successful mutation doubles with each successive developmental stage. The rate of mutation is much more rapid than encountered in natural populations: the model accelerates evolution to save computing time. A realized mutation increases $x_i(t)$ by unity. In addition, a successful change in ontogeny affects all subsequent stages; if a mutation occurs at some position i_o, then

$$x_i(t + 1) = x_i(t) + 1 \text{ for } i_o \leqslant i < N$$

A species, represented by a population of strings, loses cohesion and bifurcates to a new species under the following conditions: (1) speciation never occurs unless at least one successful mutation has occurred and (2) it is assumed that the earlier in ontogeny a successful mutation occurs, the greater the probability that this will lead to a speciation event; larger changes are more effective in destroying cohesion. Thus, the bifurcation probability $P_b(i)$ decreases with increasing i; in analogy with mutation probabilities, the model uses

$$P_b(i) = 2^{-i}$$

Given these bifurcation probabilities, it is assumed that for early bifurcations ($i < 6$) the ancestral species survives up to three bifurcation events, whereas for the later bifurcations ($i \geqslant 6$), two descendant species arise and the ancestor becomes extinct.

Descendant species are initially populations of M strings identical to that string of the subpopulation that mutated and bifurcated successfully. This is in keeping with the accelerated time frame in which reproduction is rapid enough to produce an equilibrium-level population in a single time step. The original population starts with the configuration

$$x_i(0) = 0, i = 1,2,...,N$$

When bifurcation occurs for $i \geqslant 6$, one new species is created as above, with M identical mutated strings, and the other is renumbered to label it as a daughter species of the now extinct ancestor.

The model operates as per fig. 4.4. Starting with M identical $x_i = 0$ N-tuplets (our calculations used $M = 20$, $N = 9$), one checks for each string (using a random number generator) whether a mutation takes place for $i = 1,2$, etc. . . . Should a mutation occur, x_i and its successors in that string are incremented, and a bifurcation check

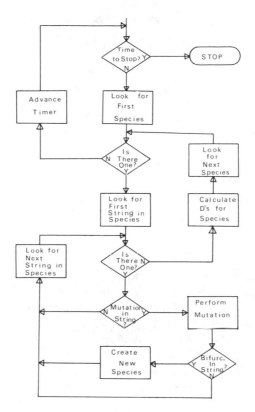

Fig. 4.4. Flow diagram of logical operations in the mutation/bifurcation model. (Redrawn from Brooks, LeBlond, and Cumming 1984.)

is made. If there is no bifurcation, one proceeds to the next string. If a bifurcation occurs for $i < 6$, a new species is created (the ancestor is destroyed when this happens for the third time) by taking away the mutated string and adding M-1 identical ones to it. For a bifurcation at $i \geq 6$, two new species are created. Each string of the species is checked for possible mutation and bifurcation. It is possible, although improbable, that two bifurcations may happen at the same time step. At each time step, all existing species are examined for mutations and bifurcations and new species stored for examination at the next time step.

The information content, and the statistical entropy associated with it, of a species is defined in terms of the number of values of x_i specifying its developmental stages, as well as in terms of their ordering. Each N-tuplet individual string is "written" in terms of an alphabet consisting of A characters ranging from the smallest x_1 found in all strings of a species up to the largest x_9:

$$A = (x_9)_{max} - (x_1)_{min} + 1$$

The number of characters of the alphabet increases with successive mutations, thereby continuously increasing the number of micro-states accessible to the system (the species).

A sequence of information measures, D_n (Gatlin 1972), charac-terizes the successively longer n-tuplet "words" (with $n \leqslant N$) within each string. The simplest measures the departure from equiproba-bility of the letters of the alphabet:

$$D_1 = \log_2 A + \sum_{j}^{A} p_j \log p_j$$

where p_j is the probability of finding $x_i = j$ for $i = 1,2,...,N$ within all the strings of the species. The summation runs from $j = (x_1)_{min}$ to $j = (x_9)_{max}$, over A letters. At each time step in the calculation process (fig. 4.4), the values of p_j are found by counting the number of $x_i = j$ for all possible i and dividing by the total number of x_i (180 for a full species from which no bifurcation has yet taken place).

The information contained in departures of pairs of successive values (x_i, x_{i+1}) from equiprobability is measured by

$$D_2 = -\sum_{j}^{A}\sum_{k}^{A} p_j p_k \log p_j p_k + \sum_{j}^{A}\sum_{k}^{A} p_j p_{jk} \log_2 p_j p_{jk}$$

where p_{jk} is the conditional probability of finding $x_{i+1} = k$ given $x_i = j$ over all values of i and all strings of a species. The probability of finding a specified pair $(x_i = j; x_{i+1} = k)$ is thus $p_j p_{jk}$. All pairs are not equally probable; the stipulation that a mutation at a certain developmental stage affects all subsequent stages means that $p_{jk} = 0$ for $k \leqslant j$. The probabilities are calculated directly at each time step by counting the number of pairs of all possible kinds over the range of the ever-growing alphabet.

Higher information measures, D_3 and D_4, defined respectively as

$$D_3 = -\sum_{j}^{A}\sum_{k}^{A}\sum_{l}^{A} p_j p_k p_l \log p_j p_k p_l$$

$$+ \sum_{j}^{A}\sum_{k}^{A}\sum_{l}^{A} p_j p_{jk} p_{jkl} \log_2 p_j p_{jk} p_{jkl}$$

$$D_4 = -\sum_j^A \sum_k^A \sum_l^A \sum_m^A P_j P_k P_l P_m \log P_j P_k P_l P_m$$

$$+ \sum_j^A \sum_k^A \sum_l^A \sum_m^A P_j P_{jk} P_{jkl} P_{jklm} \log P_j P_{jk} P_{jkl} P_{jklm}$$

were also calculated. The relative values of the various D_n may be taken as measures of the relative amounts of information contained in words of n letters in the description of the developmental code of the species. The results document the evolution of $D_1...D_4$ through time and speciation events. Expansion of the first term in D_2, D_3, D_4, allows all information measures to be written in the form

$$D_n = nH_1 - H_n^D$$

with $n \geqslant 2$ and

$$H_1 = -\sum_j^A p_j \log_2 p_j$$

$$H_n^D = -\sum_{n \text{ sums}}^A \sum^A P_{(n)} \log_2 P_{(n)}$$

where H_1 is the entropy of the single-letter occurrences in the alphabet, H_n^D is the entropy of n-tuplets, and $p_{(n)}$ is the probability of finding a specified substring of n elements in x_1.

Starting with 20 identical $x_i = 0$ 9-tuplets, the evolution and bifurcation of this original population was followed over 75 time steps. During that time, 262 species arose from the ancestral one; 193 were daughter species that followed a bifurcation, and 69 were renamed mother species from which daughter species bifurcated following a buildup of mutations late in ontogeny ($i \geqslant 6$). A partial phylogenetic tree is shown (fig. 4.5).

As mutation occurs, the alphabet's size increases and the original uniformity of the species decreases. D_1 increases on both counts (fig. 4.6), but eventually levels off to about $D_1 = 2.3$ bits; this appears to be the limit to the level of structuring, that is, of departure from equiprobability, with single-letter words under the rules given. Because A increases, the number of accessible microstates increases. There is no contradiction between the increase in information content D_1 and the simultaneous increase in entropy, because the total number of accessible microstates increases faster than the random mutations can redistribute the system among these states. The information contained in n-tuplets also increases with n; all D_n (up to $n = 4$) rapidly increase as more and more information is contained

A = SPECIES NUMBER
B = BIFURCATION HISTORY (LOCATIONS)
C = BIFURCATION HISTORY (TIMES)

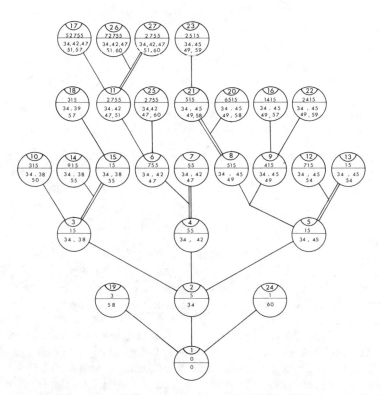

Fig. 4.5. Phylogeny of species created by the model in the first 60 time steps. Double links indicate renaming of a species following the "late-bifurcation" rule ($i \geqslant 6$). Numbers in each species indicate (1) the species' chronological order of origin (top number), (2) the developmental stages (i_s, i_{s-1},...,i_1) at which the s bifurcations occurred during phylogeny (middle number), and (3) the time steps at which these bifurcations occurred (bottom number). (Redrawn from Brooks, LeBlond, and Cumming 1984.)

in longer and longer words (fig. 4.6). Each D_n eventually levels off, so that after a long time, any increase in a species' information content occurs as an increase in the complexity of the longest words in the "dictionary." The increase in information through time occurs simultaneously with entropy increase, and the increasing concentration of information in the total ensemble indicates a nonequilibrium system according to the criteria listed in the introduction (see also Hollinger and Zenzen 1982).

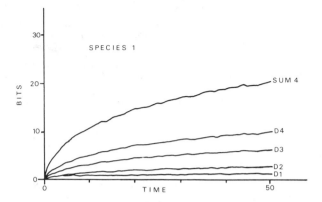

Fig. 4.6.   The evolution of the information measures $D_1...D_4$ of the ancestral species over 50 time steps. (Redrawn from Brooks, LeBlond, and Cumming 1984.)

Changes in information content upon bifurcation are of particular interest. D_1 invariably *increases* upon speciation from mother to daughter species (fig. 4.7). On the other hand, all other D_n tested *decrease* upon speciation, so that the total information content is initially lower in each of the daughter species than in its ancestor immediately prior to bifurcation. Continued mutation in all daughter species tends to bring each of their D_n into agreement with that of the ancestral species.

That D_1 takes on a larger value in the newly created species is readily understood. Consider, for example, the ancestral species configuration just before bifurcation at step 33 (table 4.1). With bifurcation at the $i = 5$ position in the twelfth string, a daughter species is created that consists of 20 identical strings (0,0,0,0,1,2,9,15,28). The alphabet of the daughter species is actually smaller ($A = 29$) than that of its mother ($A = 31$), so the increase in D_1 upon bifurcation does not arise from an increase in log A. Rather, because only a small part of the letter variability of the mother species has been transferred to its offspring, the distribution of x_i values within the alphabet is much less uniform than before (there are now only 6 letters present out of a possible 29), and the departure from randomness is therefore increased. It is the local decrease in entropy, H_1, represented by the daughter species' lower entropy state, that is responsible for the increase in information content. Or, in Gatlin's terminology, some potential information in the ancestor has been converted into stored information in the descendants.

Why do the higher information measures behave differently? When a new species is created, the number of n-tuplets that it contains

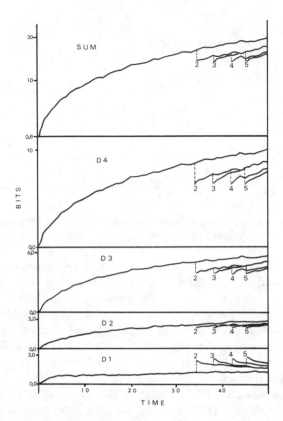

Fig. 4.7. Changes in information measures $D_1...D_4$ and their sums upon bifurcation. The lower four panels have a common scale; the scale of the top panel has been compressed. Dashed lines relate daughter to parental species. (Redrawn from Brooks, LeBlond, and Cumming 1984.)

decreases considerably when compared to those present in its ancestor. The total number of n-tuplets in an alphabet of A letters is A^n, a large number for any $n \geq 2$ at the time of the first (and later) bifurcation. Hence, although there is a sharp decrease in the n-tuplet content upon bifurcation, the ratio of actual n-tuplets appearing to all possible n-tuplets remains small in both mother and offspring species. Each population remains far from equipartition with respect to n-tuplets, and the information content upon bifurcation is dominated by the change in H_1. Hence, the decrease in D_n at bifurcation.

Gatlin assumed that we could postulate a finite alphabet for biological evolution because there was a finite number (four) of DNA bases. We consider the DNA bases to be analogous to the dots and

Table 4.1 Population and Information Changes in the First Bifurcation

	Time	Species	Population						Parent	Mutations	Bifurcations
	33	1	20						0	10	0
	0	1	2	3	5	6			10	17	24
	0	0	0	1	1	3			7	17	25
	0	0	0	0	1	1			2	8	14
	0	0	1	2	2	3			9	14	27
	0	0	0	0	1	2			5	12	21
	0	0	0	0	0	3			4	12	22
	0	0	0	0	3	4			6	9	21
A	0	0	0	0	1	3			7	13	20	$D_1 = 1.2772$ bits
	0	0	0	1	1	2			6	14	23	$D_2 = 2.4389$ bits
	0	1	1	2	2	5			10	17	22	$D_3 = 5.4087$ bits
	0	1	1	4	5	8			11	14	22	$D_4 = 8.6072$ bits
	**0	0	0	0	0	1			8	14	27
	1	1	1	2	2	3			11	15	21
	0	0	0	0	1	4			8	14	27
	0	0	0	0	1	2			7	16	27
	0	0	0	1	1	5			8	14	25
	0	0	0	0	2	6			8	12	26
	0	0	0	1	2	5			9	20	29
	0	0	0	0	1	3			7	12	22
	0	0	1	2	4	4			7	14	22
	34	1	19						0	13	1
	0	1	2	3	5	6			10	18	25
	0	0	0	1	1	3			7	17	25
	0	0	0	0	1	1			2	8	15
	0	0	1	2	2	3			9	14	28
	0	0	0	0	1	2			5	12	22
	0	0	0	0	0	3			4	12	23
	0	0	0	0	3	4			6	9	21
B	0	0	0	0	1	3			7	14	21	$D_1 = 1.2849$ bits
	0	0	0	1	1	2			6	14	23	$D_2 = 2.4395$ bits
	0	1	1	2	2	5			10	17	23	$D_3 = 5.5135$ bits
	0	1	1	4	5	8			11	14	22	$D_4 = 8.7637$ bits
	1	1	1	2	2	3			12	16	22
	0	0	0	0	1	4			8	14	27
	0	0	0	0	1	2			8	17	28
	0	0	0	1	1	5			8	14	25
	0	0	0	0	2	6			8	12	27
	0	0	0	1	2	5			9	20	30
	0	0	0	0	1	3			7	13	23
	0	0	1	2	4	4			7	14	22
C	0	0	0	0	1	2			9	15	28	$D_1 = 2.5769$ bits
												$D_2 = 2.1564$ bits
												$D_3 = 4.3215$ bits
												$D_4 = 6.5392$ bits

SOURCE: From Brooks, LeBlond, and Cummings 1984.

NOTE: (A) shows the population of 20 strings in species 1 before bifurcation. Bifurcation takes place in the twelfth string at position $i = 5$ (denoted by **). The parent population loses that string (B), and the daughter population consists of 20 strings identical to the mutated string from which bifurcation took place. Only one of these is shown C.

dashes of Morse code, and not to the letters of an alphabet at the level of populations of variable epiphenotypes. Moreover, let us suppose that whole ontogenetic programs (organisms) are the meaningful words of biological evolution, and species are evolutionary messages. Morse code is not intrinsically limited to 26 letters and 10 numerals; neither is the alphabet of biological evolution finite. We have found that if we remove the restriction of a closed alphabet and replace it with an open alphabet, then model two known dynamic mechanisms (reproduction and speciation), the result is concomitant increases in entropy and in information. This represents a modification of one of the initial assumptions of information theory and may partly explain why previous applications of information theory in biology have been less than completely satisfactory. The evolving species in the model exist in states of dynamic nonequilibrium and will remain in such states so long as the increase in maximal possible entropy outstrips the actual increase in entropy.

Entropy increase in this model results from the realization of mutations. Because the alphabet is open, mutations act as an intrinsic force driving a species away from equilibrium. Constraints on the entropy increase include faithful replication and the elimination of mutants that do not integrate with the total developmental program (i.e., canalizing selection). Those two types of constraints are represented by a fixed mutation rate and by constraints on realization of mutants such that those which had to integrate with more of the total developmental program would be less likely to succeed. Both of these notions accord with our empirical experience. In addition, we may now see that speciation also acts as a constraint on entropy increase in biological evolution. Speciation produces relatively low entropy descendant species without negentropic behavior. In this manner, biological evolution remains coherent despite its constant entropy increases.

Gatlin (1972) suggested that the major evolutionary hurdle to be crossed by living systems was that of establishing a method for efficient coding of information concerned with ontogeny and reproduction. This meant finding an optimal blend of two contrasting factors—fidelity of transmission and message variety. In most models, these factors have been considered opposing elements. One could achieve fidelity only at the expense of flexibility and, presumably, the ability to adapt to new selection pressures. Conversely, one could maintain a high degree of flexibility but only at the expense of fidelity. Mating success and complete development of offspring would become more problematical. Message fidelity is associated with D_1;

variety with $D_2...D_n$. It is possible to increase D_n without loss of fidelity if D_1 remains relatively constant throughout the process. Natural selection should heavily favor any species that stumbled across such a coding regime; Gatlin even suggested that this could serve as an objective criterion for defining the "fittest" species.

We may see by examining this model that a relatively constant D_1 is an emergent property of any evolutionary innovation that constrains entropy increase. If evolving populations and species are self-constraining, such evolutionary innovations might be analogous to the stable states of minimum entropy production in nonequilibrium thermodynamic systems. This being the case, such innovations would be favored evolutionarily just as states of minimum entropy production are favored thermodynamically. We would expect "optimal coding" to emerge universally in living systems, and for this reason alone, certain lineages will appear "selected" in retrospect (i.e., following phylogenetic analysis) because of the innovations that characterize them, not because of environmental selection continuing to "hold" them together. Innovations that constrain entropy increase might include highly faithful replication (guaranteed by the biochemical properties of nucleic acids and their associated enzymes), developmental (ensemble) constraints (Kauffman 1983), including genome homogenization (Dover 1982; Rose and Doolittle 1983) and multicellularity, sexual reproduction, organization into species, and speciation. Many of these have been recognized as significant evolutionary advances for a long time.

Layzer (1978, 1980) suggested that population genetic phenomena would be entropic and information-producing if genetic systems comprised "alpha" genes, which are expressed during ontogeny, and "beta" genes, which control recombination and mutation rates. If the "beta" genes act to constrain the actualization of genetic combinations, self-organized behavior of the form predicted by the model would arise at the population level, caused by self-organization at the genetic level (see chapter 3).

Population geneticists routinely use population variances to estimate population complexity. We hope now to show that population variance and population entropy are directly comparable measures of complexity. We make this association by considering the *redundancy* concept associated with various states of information as formalized by Gatlin (1972), using the modified notation introduced by Brooks, LeBlond, and Cumming (1984). If we can consider measures of variance to be indicators of complexity in the same sense that entropy reflects complexity, then we have a link between the theoretically conceivable and the achievable.

Redundancy at any level of information is defined as

$$RD_n = \frac{H_n^{max} - H_n^D}{H_n^{max}}$$

One transformation of this equation is

$$RD_n = 1 - \frac{H_n^D}{H_n^{max}}$$

from which we can see that Gatlin's redundancy is calculated in the same way as Landsberg's *order* (Q; see chapter 2). In the case of most population genetic studies, which are concerned with "single letter occurrences," or gene frequencies, redundancy may be defined as

$$RD_1 = 1 - \frac{H_1^D}{\log A}$$

where A = the number of different alleles in the population and $H_1^D = -\Sigma p_i \log_2 p_i$, where p_i = the frequency of the ith allele divided by the sum of the frequencies of all alleles. From this we can see that for any given set of alleles and no mutations (A is constant), an increase in H leads to a decrease in redundancy. A homogeneous population would be maximally redundant, with an H_1^D value of zero. Such a population would also have zero variance. Under such conditions, recombination would increase both the entropy and the variance of the population at the expense of its redundancy.

There is an important limitation to this level of analysis. According to the findings of the computer simulation, the smallest fraction of information present in a population will be that represented by the "single letter frequencies" or relative allele frequencies in this case, the D_1 fraction. The largest single fraction of information will be the information represented by the longest strings in the population, that is, the genotypes of organisms. For genotypes of n alleles,

$$D_n = nH_1^D - H_n^D$$

where, if we assume that each genotype is unique, $H_n^D = \log_2 N$, where N = population size. If reproduction occurs in the absence of mutations, H_n^D will increase and D_n will decrease. If mutations occur, H_1^D increases and, if the mutant allele is not spread homogeneously throughout the population, D_n can increase. The *genotypic redundancy* would be

$$RD_n = 1 - \frac{(nH_1^p - H_n^p)}{nH_1^p}$$

where nH_1^p is H_{max} for genotypes of n alleles. Thus, at the genotypic level, increasing entropy is manifested by decreasing redundancy, just as it is at the level of allele frequencies.

For a single population, redundancy would be maximal when there was a single genotype present. In this simplest case, $A = 1$, $H = 0$, and $R = 1$. Variance for such a population would be zero. As new genotypes appear through mutation and reproduction, H and A increase. Redundancy remains high so long as H remains small relative to log A; that is, so long as the actual genotypes represent a small fraction of the possible genotypes, evolutionary change will be characterized by high redundancy and low variances. If the number of actual genotypes begins to approach the maximum possible, that is, if H increases and A remains constant, evolution will be characterized by decreasing redundancy and increasing variance. In either case, evolution will be characterized by increasing entropy. In the second case, increasing variance will establish the framework within which environmental selection may play a role in evolution.

Population variances should be multivariate measures due to the number of alleles involved. The macroscopic manifestation of a genotype or an organized portion of genotype is a phenotype. Because population variance is comparable to population entropy, we think at least a crude estimate of entropy for part of the genotypic complement would be the size of the ellipse occupied by each sample when a principle component's analysis is performed on a set of phenotypic data.

It appears that estimates of population-level macroscopic behavior will be inadequate if based only on allele frequency data. Only the smallest fraction of a population's information content can be measured from allele frequency data alone, leading to an underestimation of the population's inherent organization and complexity. Nonetheless, allele frequencies are necessary for estimating the higher level information content of any population.

Information Partitioning and Multiple Populations

When we consider more than a single population, we must consider between-population differences as well as within-population differences. A species may be complex in two ways. First, it may be polymorphic. If so, the complexity of information may be high within a population but between-population differences may be low. Second, it may be polytypic. If so, the complexity of information may be low within a population but between-population differences may be high.

If the investigator works with only a few loci over two or more populations, he or she can calculate H values for each population and test for differences by calculating a variance for each. Variances may be calculated, following Basharin (1959) and Lloyd, Zar, and Karr (1968), thusly:

$$S_H^2 = \frac{p_i \log_2 p_i - (p_i \log_2 p_i)^2/N}{N^2}$$

where S_H^2 is the variance and N is population size (see Zar 1984). Differences between two populations can be tested using a t-test proposed by Hutcheson (1970) where

$$t = \frac{H_1 - H_2}{S_{H_1 - H_2}}$$

$$S_{H_1 - H_2} = S_{H_1}^2 + S_{H_2}^2$$

and the degrees of freedom are approximated by

$$df = \frac{(S_{H_1}^2 + S_{H_2}^2)^2}{\dfrac{(S_{H_1}^2)^2}{N_1} + \dfrac{(S_{H_2}^2)^2}{N_2}}$$

For many characters the investigator could partition the variance and test for differences using standard techniques. However, for multivariate analysis one needs to assume multivariate normality to do so. One reason we cannot presently equate variance with entropy exactly is because we do not know if assumptions of multivariate normality are warranted.

The rate of change in groups of populations depends on factors that affect the rate of change within populations (is the new allele or phenotype selected for?) and the migration rate between populations. Obviously, the degree of isolation of populations is directly correlated with the rate at which a favored mutation can spread.

Patterns of Change in Information

We can summarize the entropic changes in information that occur in populations by considering the possible patterns of reproduction

Table 4.2 Four Classes of Linkage Patterns between New Phenotype and Ancestral Phenotype

	New pattern established and maintained (+)	New pattern not established or maintained (−)
Old pattern maintained (+)	Case I (+/+)	Case II (+/−)
Old pattern lost (−)	Case III (−/+)	Case IV (−/−)

that might emerge. We suggested that four outcomes were possible, depending on whether the old reproductive pattern (that between ancestral phenotypes) and/or the new reproductive pattern (between new phenotypes and/or old and new phenotypes) were maintained or established and maintained (Wiley and Brooks 1982). In table 4.2, the four expected outcomes are summarized as cases I to IV. These outcomes are translated into population-level phenomena in table 4.3. We will see that the outcomes generalized in table 4.2 can also be translated into speciation outcomes (see table 4.7). Polymorphism (case I) results when both phenotypes are found in the same deme. A polymorphic species is one in which the same polymorphisms are found in all populations. Further, the characters involved are not correlated with the sex of the individual (sexual dimorphism) nor with changes in ontogeny. Geographical variation (also case I) obtains when there is a correlation between increasing complexity and the geographical distribution of different phenotypes. Species geo-

Table 4.3 Four Classes of Linkage Pattern Changes When Information Changes are Under a Critical Value which would Cause Disruption of Cross Linkages

	New pattern established and maintained	New pattern not established or maintained
Old pattern maintained	I. Polymorphism and/or geographical variation	II. Canalizing selection
Old pattern lost	III. Anagenesis via directional selection	IV. Not realized

graphical variation is a phenomenon in which the species as a whole is complex, whereas each deme is relatively simple. Geographical variation therefore represents an increase in complexity at the species level without a parallel increase to the same level of complexity in any one deme.

Stable polymorphisms are established when the reproductive linkage patterns between phenotypes are equally probable (no selection against heterozygotes), when heterozygotes are selected for, or when gene frequency-dependent selection is operating. If neutrality obtains, a polymorphic species would be relatively disordered in terms of information but highly organized in terms of cohesion as a monotypic species. If selection obtains, the species would be disorganized to a degree directly related to the selection pressures that obtain.

Species occupying a larger geographical range than the cruising range of any one individual are less cohesive, as discussed below, and thus show an inherent degree of disorganization. In such a case, an increase in information complexity may not cause a significant increase in the disorganization of the species as a whole, even with selection.

Case II represents canalizing selection in which the mutation does not become established because it results in death, sterility, or reduced fecundity. Fixation can occur via drift if the change lowers fitness without causing sterility or death.

Case III represents anagenesis, and we have discussed its dynamics earlier in the chapter. The outcome of anagenesis is the fixation of a new phenotype. If the new phenotype is not inherently more complex (e.g., substitution of one allele for another), the outcome does not result in an overall increase in entropy. If the new phenotype is more complex (e.g., an addition to the ontogenetic program), then the outcome results in an overall increase in the entropy of the lineage. The anagenetic entropy changes are summarized thus (Brooks and Wiley 1984):

$$\text{Substitution: } dS_l = S_a - S_p = 0$$
$$\text{Addition: } \quad dS_l = S_a - S_p > 0$$

where dS_l is the change of the entropy of information of the lineage (= species), S_a is the entropy of the new (apomorphic, novel) phenotype, and S_p is the entropy of the old (plesiomorphic, ancestral) phenotype. Finally, we have suggested that case IV is not realized within species because such changes would be eliminated under case II (Wiley and Brooks 1982; Brooks and Wiley 1984).

Proximal Mechanisms of Changes in Information

We termed *proximal mechanisms of change* those mechanisms that promote or inhibit the spread of evolutionary novelties (Brooks and Wiley 1984) to distinguish these mechanisms from *ultimate mechanisms of change,* mutations (genetic and chromosomal), and cytoplasmic reorganizations. Four such proximal mechanisms are listed below.

1. *Canalizing selection.* Any evolutionary novelty must be minimally accommodated with the rest of the genome of an individual for that individual to grow and reproduce. Canalizing selection provides the developmental constraints on the kinds of changes that can occur in individuals and thus is the basis for historical constraints. Because mutations are not equally likely to be tolerated at every locus (especially those involved in gene regulation), we predict that there is an inherent bias in the direction evolution can proceed from any given starting point. Such a bias would exist regardless of the randomness of mutations over all loci. Canalizing selection also provides constraints on the level of functional complexes. The whole effect, as Lauder (1982) has deduced, produces an intrinsic phylogenetic component of ontogeny and function that "may severely constrain the directions of structural modification that can occur." Thus canalizing selection is deterministic and can limit the kinds of variation that can occur. This is what we think Richard Goldschmidt meant by "orthogenesis," an "orthogenesis" devoid of teleology and vitalism. This is what Alberch (1982) meant by the constraints ontogeny imposes on phylogenetic change.

2. *Directional selection.* A new phenotype will replace an old one if the new phenotype has an advantage in reproductive success. This may occur because the new phenotype is more fecund or because it outcompetes the old one. Directional selection may be periodic, favoring one phenotype during one season, frequency dependent, or geographically manifested. The result is change in the frequency of phenotypes and genotypes, resulting in polymorphisms, clines, or complete replacement. The significance various investigators place on these phenomena depends on the importance they attach to the role of directional selection in the evolutionary process and not, to our minds, the reality of these phenomena. Is evolution primarily adaptation to the external environment? If so, natural selection may be viewed as a creative process (Simpson 1947; Mayr 1963) and a process that gives direction to evolution (see Mayr 1982). As Alberch (1982) recently stated: "In this approach environmental factors are

the driving force in evolutionary change, since they define the selective pressures associated with different phenotypes. In this case, adaptation and morphological diversification are one and the same process. Organisms diversify by becoming adapted to different environments." Of course, not even the strongest proponent of environmental selection (frequently termed simply *natural selection*) would suggest that, say, drift never occurred or that developmental constraints did not play a part in the process. And, we cannot imagine the most militant anti-Darwinian claiming that the correlation between melanistic insects and soot-covered trees has nothing to do with environmental selection or that polar bears have white pelts purely as a matter of happenstance. In a real sense, one's views about this controversy depend on how one views variation. If there are no limits on variation in a lineage, then environmental selection is, potentially, an omnipresent and omnipotent force. If variation is constrained by history and its manifestation, development, then directional selection is itself constrained. We take the latter view. The questions then revolve around history. In the phylogeny of a clade, how many speciation events are correlated with changes in behavior or environment of great enough magnitude to force irreversible changes other than those inherent trends? The extent to which we observe such phenomena will be the extent to which the evolution of the species is correlated with such changes. And only when there is such a correlation may we ask the question of environmental causation. This would involve determining whether the changes that occur in the descendant(s) of an ancestor are themselves correlated with the environmental change or if they may be correlated with another factor (which may also cause directional change) operating during speciation. In this view, environmental selection is something we test for empirically rather than something assumed a priori.

3. *Drift*. A new genotype or phenotype may replace an ancestral one by chance (Wright 1931). This mechanism is attractive because it can explain changes in populations that show no correlation between genotypes and/or phenotypes and the environment. It is also attractive in explaining the fixation of a maladapted trait. Its major drawback as a general mechanism is that it works most effectively only in populations with small effective breeding sizes.

4. *Concerted evolution*. A new phenotype may replace an old phenotype because of the wholesale conversion of a part of the gene pool via molecular drive (see chapter 3). The widespread occurrence of repetitive DNA sequences argues that molecular drive is a common phenomenon. As a postulated mode of speciation, it is attractive

because (1) it involves an irreversible mechanism, (2) it works on populations of all sizes, and (3) speciation does not have to be correlated with environmental differences. In addition, since this process involves the production and converted turnover of repetitive DNA sequences, it is an evolutionary mechanism that operates in conjunction with a high level of redundancy. Thus, according to our reasoning, it should be a low-variance, low-entropy mode of change. Dover (1982) and Ohta and Dover (1984) have stressed that molecular drive should result in low-variance changes in populations.

Note that canalizing selection and concerted evolution are, along with reproduction, inherently irreversible processes. Their operation should be characterized by constrained (organized) increases in entropy. Environmental selection and drift appeal to probabilistic responses to, or in spite of, extrinsic factors to explain macroscopic organization. To the extent that such extrinsic phenomena occur, they will have an impact on the evolution of any given population. However, since there are inherently irreversible processes operating continuously in populations, we expect the effects of the extrinsic factors to be more in the nature of functional constraints on entropic behavior rather than determinants of such behavior. In this sense, we are decidedly not "selectionists." Rather than assuming that environmental selection (extrinsic factors broadly construed) is the "only direction giving force" in evolution (Mayr 1982), we do not think it is direction giving at all. Nonetheless, it is a real and important phenomenon.

Cohesion

When asexual and sexual reproduction are compared microscopically, it can be seen that (1) both modes increase the variety of genotypes (tend to randomize information states) in a population and that (2) sexual reproduction provides higher rates of information randomization through enhanced recombination. The higher rate should provide sexually reproducing species with greater probabilities of having successful genotypes available in varying environmental regimes (Maynard Smith, 1978). On the other hand, production of gametes involves partial fragmentation of each genetic combination. This would make it harder to establish and maintain optimal genotypes in stable environments. In addition, each parent contributes only 50% of the genetic component to each offspring and expends energy finding a mate, which could have been converted into offspring. Thus, there appears to be a negative "cost" associated with sexual reproduction (William, 1975). This raises an important paradox: if sexual reproduction is "costly," why it is so prevalent among protists, plants, and animals?

We believe this paradox can be resolved by examining the *macroscopic* aspects of sexual reproduction, which we call *cohesion*. Sexual reproduction has two important macroscopic effects. First, as we mentioned in our discussion of the computer simulation model, entropy increases in populations will be minimized by any mechanism that increases genetic homogenization. In order for a mutation to become fixed throughout a population of asexual organisms, the mutation must arise multiple times or the nonmutant members of the population must all die. Even a combination of these processes is likely to be slower than sexual reproduction in spreading mutants through a population. Second, sexually reproducing species are able to maintain their coherence as functional wholes despite increased variation (increasing entropy) better than asexually reproducing species. Gametic recombination is an excellent mechanism for maintaining the status quo. Asexual species are more easily fragmented into multiple independent systems by mutations.

Cohesion thus has a constraining, organizing influence on information changes in populations. We will show that this influence is capable of producing the curves we have shown. There is no cost associated with sexual reproduction when it is viewed from a macroscopic perspective. If, as we suggest, evolution is a macroscopic phenomenon, the success of sexual reproduction is due to the effects of cohesion.

Cohesion can be viewed in two very different lights. First, it could be considered a force opposing the entropic behavior of evolution. Organized species and populations would result from this interaction of opposing forces. Cohesion as a quantified function would then exhibit negentropic behavior. We will argue an alternative view, namely, that cohesion, like information, behaves entropically. Organized species and populations emerge from the complementary interaction of these two coupled phenomena. There is no need to postulate negentropic behavior to obtain organization.

Our reasoning is this: a common deterministic outcome of reproduction is the dispersal of descendant organisms beyond the immediate vicinity of the parent(s). This dispersal is constrained proximally by the functional capabilities and habitat requirements of particular organisms and is constrained ultimately by the necessity of finding a mate. That is, members of any given species may sometimes find themselves far from other members of the species and in unusual habitats, but they will never be found in established populations unless they find mates every generation. We would expect a population to exhibit a deterministic trend toward expanding its geographical range, all other things being equal. This in turn would lead to an attenuation of cohesion among members of a species. Such attenuation can be imposed by extrinsic factors such as geographical or climatic alteration as well. We will show that this attenuation of cohesion is accompanied by an increase in entropy and thus represents a dissipation of cohesion.

A useful analogy would be molecular diffusion. Diffusion is an entropic process, and Hollinger and Zenzen (1982) have reminded us that diffusion is *slowed,* not *caused,* by molecular collisions. In an analogous manner, population dispersion is an entropic phenomenon slowed by the requirements of mate-finding and reproduction.

Cohesion is one of the more difficult concepts with which we have attempted to deal because, like any other physical capacity involving interactions, it defies easy characterization and quantification. We began thinking about cohesion with the idea that species were "held together" as coherent systems by past history and that at least some sexually or parasexually reproducing species were held together by gene flow, that is, reproductive linkages among the populations comprising the species, as well. Asexual species and allopatric populations of sexual species, while having historical continuity, would lack cohesion. Many factors can affect cohesion, and this will make quantification of the concept difficult, because it will depend on quantifying overall rates of gene flow among populations, a measure that has eluded population geneticists to date. However, we think that such a capacity as cohesion exists; populations exist through time only because they perpetuate themselves by reproduction. Speciation occurs, indicating that lineages split, become noncohesive, and evolve independently. Replicate geographical distribution patterns emerge because gene flow is cut off in many different species at the same time, indicating that cohesion is an important component of a species' makeup.

Factors Affecting Cohesion

1. *Sexual systems.* The maximally cohesive sexual system would be one promoting panmixis. Maximum cohesion could be achieved, all other things being equal, only in the case of hermaphrodites which both self- and cross-fertilize. Obligately selfing sexual species and

asexual species exhibit no cohesion. Species with discrete sexes occupy the full range of intermediate degrees of cohesion.

2. *Panmixis*. If all members of a population have an equal probability of mating with any other member of that population, the population may be considered panmictic. A panmictic population will be more cohesive than one displaying assortative mating in which the probability of mating varies with increased or decreased similarity. True panmixis can occur only when all individuals are within cruising range of each other. Species that occupy ranges larger than the cruising range of any one individual cannot be strictly panmictic because the probability of mating is not equal among all members of the species. Rather, the probability of mating is relatively high for individuals with broadly overlapping cruising ranges, relatively low for individuals with small overlap, and nil for individuals whose cruising ranges do not overlap. For a sexually reproducing population of outcrossing hermaphrodites, one measure of panmixis might be

$$p(M) = N\left(\frac{1}{N_i}\right)$$

where $p(M)$ is the probability of mating for a selected individual, N is the number of individuals, and N_i is the ith individual. If panmixis obtains, the probability that our selected individual will mate with any other individual or with itself is equal, and $p(M)$ has a variance (σ^2) of zero. Obviously, if different probabilities obtain, we may express them as a collection of unequal probabilities whose variance will be greater than zero.

While panmixis is a component of cohesion, the two terms are not synonymous. This can be demonstrated by considering the relationship between increases in the number of individuals and the effects of such increases on measures of panmixis. If we begin with a population of obligate hermaphrodites inhabiting a range equal to the vagility of each individual, and if we assume that the probability of self-fertilization is the same as the probability of mating with another individual, then perfect panmixis obtains:

$$p(M) = N\left(\frac{1}{N_i}\right)$$
$$= 8(1/8)$$
$$= 1$$

We could measure the "entropy" of such a system using the following measure:

$$H = -\sum (p_i \log_2 p_i)$$
$$= -8\left(\frac{1}{8} \log_2 \frac{1}{8}\right)$$
$$= -3 \text{ bits}$$

If we increase the number of individuals to sixteen, we can obtain a second measure:

$$H = -\sum (p_i \log_2 p_i)$$
$$= -16\left(\frac{1}{16} \log_2 \frac{1}{16}\right)$$
$$= 4 \text{ bits}$$

In other words, if the entropy of the population was measured by a measure of mating probability (or an average if these probabilities are not equal), the entropy of cohesion increases with increasing numbers of individuals. Yet, can we really say that our population of sixteen individuals is less cohesive than our population of eight individuals? We think not. There is more information but the same degree of cohesion. Indeed, there is every reason to think that the opposite is true since we have the intuitive idea that more reproductive activity means more cohesion and since we know that by this measure the population would be most "cohesive" just before it went extinct. But, if we consider the variance of our measure of panmixis to be relevant, we see a different picture. The variance is zero in both cases (i.e., $N = 8$, $N = 16$), indicating that, all other factors being equal, there has been no change in the cohesion of the population with increasing numbers of individuals. This is our clue that the variance of panmixis is an indicator of entropic behavior for cohesion.

3. *Deterministic factors.* The most obvious deterministic factor that might affect cohesion is the presence of two or more kinds of individuals, that is, two or more genotypes. A population with several different genotypes will be less than maximally cohesive if one or more genotype is selected against. In archetypal ("good") biological species, cohesion between two different species is zero because of premating and/or postmating isolating mechanisms (Dobzhansky 1970). Two not-so-obvious deterministic factors that affect cohesion are dispersion within the range of the species and

the extent of the range relative to the vagility of its members. Individuals of a species are rarely, if ever, dispersed equally within the range because the extent to which the available habitat is clumped is a deterministic factor relative to the dispersion of individuals. If "gene flow" within populations exceeds gene flow between populations (*sensu* Endler 1977), the species is not panmictic and the level of cohesion is relatively low. Of course, other deterministic factors may counter density and dispersion effects. These include such things as mate-attraction mechanisms (pheromones, etc.) or migration to particular breeding sites (effectively increasing density). Thus, when we discuss density factors in subsequent sections, we are assuming that these other deterministic factors are not present as a simplifying assumption.

4. *Stochastic factors*. Cohesion may be raised or lowered by chance factors that affect the vagility of individuals. Additionally, density may decrease to the point where chance alone may further decrease cohesion simply because individuals do not meet frequently enough to ensure that the lineage can continue.

5. *Fecundity and survivorship*. Increases in number of individuals may be a function of increased rates of reproduction or of increased survivorship. However, increased survivorship will result in increased reproduction in the next generation (all other factors equal); so for our purposes they amount to the same thing one generation removed.

We suggest that there are two levels at which cohesion must be considered: (1) within populations and (2) among populations. In the following sections we will consider factors that promote and inhibit cohesion at each of these levels.

Cohesion Within a Population

We may think of populations, or demes, as being of two sorts: closed populations that do not share mating bonds with other populations and open populations linked to other populations through reproduction. Closed populations are more cohesive than open populations because the phenomenon of migration itself results in an increase in the variance of panmixis (i.e., the migrants have a zero probability of mating with nonmigrants of their own deme and immigrants possess lower mating probabilities than residents). Thus, within-population cohesion is *inversely* proportional to among-population cohesion (all other factors being equal). High within-population cohesion means low among-population cohesion.

What factors might affect within-population cohesion? Three obvious factors are vagility, geographical range, and demography. Cohesion will be high if vagility is high, low in populations whose geographical area is larger than the vagility of members of the population, and low if the population is composed of clumped subpopulations.

Another factor is differences between phenotypes constituting the population. If matings between different phenotypes result in offspring with mating probabilities equal to or exceeding the mating probabilities of parental phenotypes, then these differences do not affect or may actually promote cohesion. If a particular phenotype is selected against, the cohesion of the population as a whole might remain constant, and the elimination of the phenotype via natural selection can be viewed as a mechanism that promotes cohesion of the population as a whole. If the hybrids of between-phenotype matings are eliminated but there is no selection against the original phenotypes themselves, the elimination of hybrids promotes a decrease in cohesion and any tendency toward assortative mating will result in a schism within the population evidenced by an increase in *within*-phenotype cohesion and a decrease in *among*-phenotype cohesion. In short, cohesion within populations is promoted and inhibited by processes elucidated over the years by population geneticists and population biologists. The results of these processes may be measured in a variety of ways (calculating selection coefficients, measuring gene flow, calculating panmixis measures and variances).

Cohesion Among Populations

Except for some types of sympatric speciation, the origin of species is thought to involve not single populations but several populations spread over a geographical landscape. In this section, we will consider measures of cohesion and measures of cohesion entropy between populations that reflect the cohesion or lack thereof of species.

We will use "deme" to refer to the local population and "population" ambiguously to refer to one or more demes. A "system" is a series of reproductively connected demes. Demes may be well defined (low variance of panmixis, little migration) or diffuse (high variance of panmixis, high migration rates, high levels of gene flow, less isolation). Obviously, a species that is panmictic is not comprised of demes but of a single population. However, a species that is not panmictic may not be composed of discrete or semidiscrete demes but simply a single large population in which the variance of panmixis is high because of distance. We restrict our present discussion to species composed of more or less discrete demes.

Consider a species comprised of only two demes. Two cohesion measures may be calculated, cohesion within each deme and cohesion between demes. The extent to which the species displays cohesion at all will not depend on within-deme cohesion but between-deme cohesion. In other words, if gene flow is nil, the species is not a cohesive system but two separate, isolated systems. Our task, then, is to suggest a measure or measures of between-deme cohesion.

One possible measure would be to measure the Wahlund effect. As pointed out by Nei (1965), Wahlund (1928) showed that in a species divided into many small demes, each deme being a random mating system, "the frequency of heterozygotes for a locus with two alleles decreases by an amount equal to twice the variance of gene frequencies among subpopulations compared with that expected in a single random mating population, while the frequency of homozygotes increase by the same amount." Measurements utilizing F statistics have been developed by Wright (1969, 1978) and extended to multiple allele systems by Nei (1965). Another possible measure would be a measure of gene flow. Slatkin (1981) proposed a measure, the *conditional average frequency*, which is relatively independent of selection intensities and mutation rates but highly dependent on the overall level of gene flow between demes. Unfortunately, Slatkin's measure cannot be used to estimate migration rates, and thus actual gene flow measurements remain elusive. However, given the appropriate assumptions, his measure can be used to distinguish low, intermediate, and high gene flow levels.

Cohesion and Entropy

We have suggested that increasing isolation of demes and lineage splits are entropic phenomena (Wiley and Brooks 1982, 1983). We arrived at this deduction intuitively by considering the consequences of making a cohesive reproductive system into two separate and noncohesive reproductive systems. In this section, we will suggest one possible way to quantify our intuition.

Cohesion among demes is promoted by gene flow (successful migrations and subsequent reproduction of immigrants). High levels of cohesion should be correlated with low states of cohesion entropy.

Isolation of demes or groups of demes is promoted by low to nil levels of gene flow. Going from a relatively high level of cohesion to a low level of cohesion should be correlated with high levels of cohesion entropy promoted by low levels of gene flow, resulting in increased levels of isolation.

As discussed by Shields (1982), the term *inbreeding* may mean several things. For example, many authors define inbreeding in such a way that a single panmictic population is, by definition, *outbred* (see Crow and Kimura 1970), even though all of the individuals may share a pattern of ancestry not shared by members of other populations of the species. Such a definition is not useful for our purposes. We will use the word *isolated* to denote a population that shares no reproductive ties with other populations.

There are two factors to consider in measuring the entropy level of cohesion of a species. First, we must consider the demography of the demes or populations and the reproductive linkages they share. Second, we must consider the degree to which each deme is isolated.

The demography and reproductive linkages of demes within a species combine to form a between-deme topology. If we know which demes are linked by reproductive ties (i.e., which demes share genes), we can calculate a statistical entropy measure for the species as a whole using the formula

$$H_c = -\sum p_i \log_2 p_i$$

where H_c is the statistical entropy of the ensemble and p_i is the configuration probability of the ith deme. For example, a number of such measures are calculated for the five populations of the species *Xus yus* shown in fig. 4.8 as its demes drop reproductive linkages through time.

For fig. 4.8*a:*

$$H_c = -(5)(4/20 \log_2 4/20)$$
$$= 2.322 \text{ bits}$$

For fig. 4.8*b:*

$$H_c = -(2)(3/12 \log_2 3/12) - (3)(2/12 \log_2 2/12)$$
$$= 2.292 \text{ bits}$$

For fig. 4.8*c:*

$$H_c = -(3)(2/10 \log_2 2/10) - (3/10 \log_2 3/10) - (1/10 \log_2 1/10)$$
$$= 2.246 \text{ bits}$$

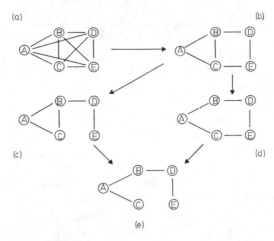

Fig. 4.8. Changing reproductive linkages of the hypothetical species *Xus yus* through time. *a* = Initial configuration of reproductive linkages among populations *A* to *E*; *b* = first pattern of change in reproductive linkage patterns; *c* and *d* = alternative changes from *b* with same number of reproductive links among populations; *e* = final configuration. There are a number of alternative configurations for *c* and *e*, which have identical entropy values.

For fig. 4.8*d:*

$$H_c = -(5)(2/10 \log_2 2/10)$$
$$= 2.322 \text{ bits}$$

For fig. 4.8*e:*

$$H_c = -(3)(2/8 \log_2 2/8) - (2)(1/8 \log_2 1/8)$$
$$= 2.250 \text{ bits}$$

If we graph these changes through time (fig. 4.9), we can see that the species' cohesion entropy level fluctuates according to the relative number of connections among populations. These values alone cannot provide an adequate entropy measure since we would expect qualitatively that the cohesion entropy of *Xus yus* would increase with decreased linkages rather than fluctuate within definable boundaries. This is the type of behavior we would expect if cohesion were treated as a boundary conditions effect rather than an initial conditions effect. Because sexual reproduction involves two separate organisms, there is a sense in which cohesion is "extrinsic" to organisms, while still being part of the initial conditions provided by ontogeny. Thus, cohesion produces a form of *self-imposed organization* at the level of populations and species in a manner anal-

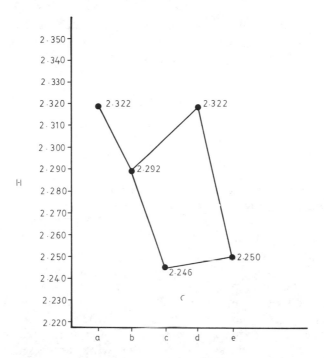

Fig. 4.9. Entropy values for cohesion changes among populations, as shown in fig. 4.8, when entropy is calculated without taking inbreeding into account. Note that entropy values fluctuate within set boundaries.

ogous to the self-imposed organization resulting from epigenetic phenomena in ontogeny.

   We suggest that what is missing is a measure of the degree of isolation of the populations provided by an estimate of "inbreeding." From population genetic theory, the probability that two alleles chosen at random within the same population are identical by descent is termed F. The balance between inbreeding and outbreeding due to successful migration produces a value, F, such that

$$F = \frac{(1 - M)^2}{1N - (2N - 1) - (1 - M)^2}$$

which can be simplified, given low rates of migration, to

$$F = \frac{1}{4NM + 1}$$

where N is the population size and M is the migration rate. Such a measure has one characteristic suitable for our purposes. If the migration rate between demes is zero, then

$$F = \frac{1}{4N(0) + 1} = 1$$

If the migration rate between "demes" is 1, then

$$F = \frac{(1 - 1)}{2N - (2N - 1)(1 - 1)^2} = 0$$

and there is only a single population with no deme structure. However, it has one characteristic that is not suitable for our purposes. At intermediate levels of gene flow, it is influenced by the number of individuals within a population. Yet, if we have a true measure of isolation (contra inbreeding), we would expect the size of populations to have little influence on cohesion relative to migration (unless migration were a function of population size). We believe we can get around this problem to illustrate some points by holding population sizes equal and constant in all populations.

The preceding formulas have some characteristics that may not be suitable for our purpose. For one thing, they are based on "the island model" population structure discussed by Wright (1968, 1978), which assumes that the migrants are representative of the species as a whole. Thus, if we attempted to deal with information changes (gene frequency changes) as well as cohesion changes, the results might not be realistic as generalizations. However, the cohesion changes we wish to explore are independent of differences in alleles and dependent entirely on gene flow under the assumption of neutrality (the "simplest" assumption we can make to isolate our considerations of cohesion from the impact of changes in information). Thus, for our immediate purpose this measure should do, although it will not be suitable for a full integration of cohesion and information. That is, it might not be the most appropriate measure when we get to the point of investigating the covariance of cohesion and information. Instead, the measure of isolation will depend on the population structure of the species investigated, as discussed by Wright (1969).

In spite of these problems we believe we can use this measure to illustrate a point—that the level of cohesion entropy increases with increasing isolation. We suggest that

$$H_c = -\sum F_i(p_i \log_2 p_i)$$

and that F_i is the cohesion analog to "k" in statistical mechanics, where

$$S = -k \sum p_i \ln p_i$$

One obvious difference, of course, is that k is a constant, frequently Boltzmann's constant, where F_i may vary among demes. In special cases for which all F values are equal, we may write our general equation as

$$H_c = -F \sum p_i \log_2 p_i$$

Furthermore, an average F, which we might call $\bar{F}$, can be calculated according to

$$\bar{F} = \frac{F_i}{n}$$

where n is the number of populations, and a general approximation would be

$$H_c = -\bar{F} \sum p_i \log_2 p_i$$

Table 4.4 contains the statistical entropy values calculated as above for the history of *Xus yus* as depicted in fig. 4.8. Each population has been assigned a migration rate (M) and a population size (N) from which F values are calculated. From these values, cohesion entropy values, H_c, are calculated using our general formula. Note that loss of cohesion is associated with increasing cohesion entropy.

The model on which the calculations presented in table 4.4 are based supposes that dropping reproductive linkages between populations increases the isolation of these populations. In a sense, this is a model based on deterministic gene flow. That is, it is a model that supposes cutting off gene flow between A and E does not increase gene flow between, say, A and B. In such a case, the level of cohesion entropy goes up as reproductive linkages are broken, as our theory would imply given the model (table 4.4). When we examine the alternate case, in which cutting off gene flow between certain populations does not affect their level of isolation (table 4.5), the species fluctuates in exactly the same manner as the topological information value. One interpretation is that a species' cohesion is relatively unaffected by how genes are shared.

We may now extend our thinking to the species that becomes noncohesive. We may define a noncohesive species as a species with at least two allopatric populations, that is, two populations that do

Table 4.4 Levels of Cohesion Entropy for Hypothetical Species *Xus yus* through Time

Configuration	Population	N	m	F	$-p_i \log_2 p_i$	$-Fp_i \log_2 p_i$
a	A	100	.01	.20	.4644	.09288
	B	100	.01	.20	.4644	.09288
	C	100	.01	.20	.4644	.09288
	D	100	.01	.20	.4644	.09288
	E	100	.01	.20	.4644	.09288
				$\bar{F} = .20$	$2.322 =$ $-\sum p_i \log_2 p_i$	$H_c = .4644$
b	A	100	.008	.2381	.4308	.1025
	B	100	.009	.2174	.500	.1087
	C	100	.009	.2174	.500	.1087
	D	100	.008	.2381	.4308	.1025
	E	100	.008	.2381	.4308	.1025
				$\bar{F} = .2298$	$2.2524 =$ $-\sum p_i \log_2 p_i$	$H_c = .5247$
c	A	100	.008	.2381	.4643	.1105
	B	100	.009	.2174	.5211	.1133
	C	100	.008	.2381	.5211	.1240
	D	100	.008	.2381	.5211	.1241
	E	100	.004	.3846	.3322	.1278
				$\bar{F} = .2633$	$2.3598 =$ $-\sum p_i \log_2 p_i$	$H_c = .5997$
d	A	100	.008	.2381	.4644	.1106
	B	100	.008	.2381	.4644	.1106
	C	100	.008	.2381	.4644	.1106
	D	100	.008	.2381	.4644	.1106
	E	100	.008	.2381	.4644	.1106
				$\bar{F} = .2381$	$2.322 =$ $-\sum p_i \log_2 p_i$	$H_c = .5530$
e	A	100	.008	.2381	.500	.1191
	B	100	.008	.2386	.500	.1191
	C	100	.004	.3846	.375	.1442
	D	100	.008	.2381	.500	.1191
	E	100	.004	.2346	.375	.1442
				$\bar{F} = .2967$	$2.250 =$ $-\sum p_i \log_2 p_i$	$H_c = .6457$

NOTE: Entropy calculated from topology of breeding connections among populations (as shown in fig. 4.8) and various levels of isolation. In this series of calculations, it is assumed that losing reproductive linkages decreases migration, thus increasing "inbreeding".

Table 4.5 Levels of Cohesion Entropy for *Xus yus* through Time as Reflected by the Topology of Populations (fig. 4.8), Assuming Overall Migration Rates Remain Constant

Configuration	$F (= \bar{F})$	$-\sum p_i \log_2 p_i$	H_c
a	.2	2.3220	.4644
b	.2	2.2924	.4584
c	.2	2.3598	.4719
d	.2	2.3220	.4644
e	.2	2.2500	.4500

NOTE: All data for populations (except *m*, which is 0.1 for all cases) are the same as in table 4.4. This scenario is not based on the assumption that "inbreeding" increases with increasing isolation; thus, *F* cannot be calculated by the formula we have discussed earlier. We simply picked $F = 0.2$ to make the initial calculations equivalent.

not share genes. Each population may be composed of one or more demes. For example, *Xus yus* might become noncohesive in a manner detailed in fig. 4.10. We would have to go from a single system of five demes to two systems, one comprising three demes and the other comprising two demes. The migration values for each population are given in the figure, and each is composed of 100 individuals at any one time. Further, each lost linkage is assumed to increase isolation for that population which loses the linkage. Given this, the overall entropy levels of cohesion and a graph of these levels for *within systems* is given in fig. 4.11. The total entropy of the final result, shown in fig. 4.10*d*, indicates a drastic reduction of entropy levels within the new subsystems (ABC and DE, respectively), but the overall entropy, $H_{cABC} + H_{cDE}$ has increased since $H_{ct} = .5282$.

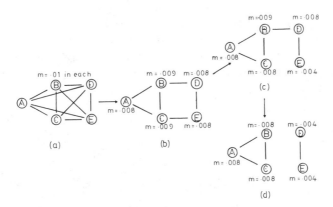

Fig. 4.10. Another possible history of cohesion changes in the species *Xus yus*. *m* = Migration rate.

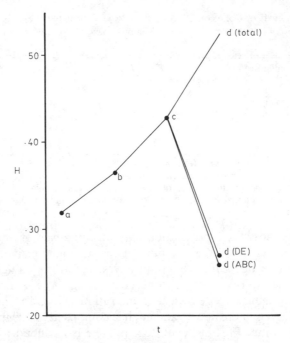

Fig. 4.11. Entropy changes associated with cohesion changes in species *Xus yus* calculated on the basis of migration rates and deme topological structures shown in fig. 4.10.

In terms of cohesion, there are now two separate systems connected to each other by historical links only. The between-lineage cohesion is nil, meaning that the isolation coefficient for each lineage, relative to the other, is $F = 1$. Thus, the entropy of the ensemble of two independent systems is equal to that suggested by the topology shown in fig. 4.12, where the two independent systems are shown connected to each other by their historical membership in the ancestral group of populations, represented as a tree diagram or directed graph. Thus, since

$$H_c = -\sum F_i(p_i \log_2 p_i)$$

Then

$$H_c = -\sum p_i \log_2 p_i$$

when $F = 1$. In the simplest case, shown in fig. 4.12:

$$H_c = -(1/2 \log_2 1/2) - (2)(1/4 \log_2 1/4)$$
$$= 1.5 \text{ bits}$$

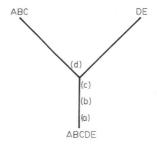

Fig. 4.12. Graphic representation of partitioning of ancestral species *ABCDE* in fig. 4.10 into descendant species *ABC* and *DE*. *a–d* refer to different configurations in fig. 4.10.

Notice that when $F = 1$, we move from one functional level to another. In chapter 3, we considered an analogous situation when moving from the level of regulatory networks to cell lineages. In the present case, we find that when moving from the level of one group of populations to two (or more) separate groups, the entropy of the ensemble is dominated by the independence of the systems, and resolution of the cohesion within each group of populations is lost. However, as in previous sections, shifts from one functional level to another may not preserve absolute magnitudes of entropy due to coarse graining, but do preserve relative entropic behavior of the systems. When such changes in cohesion are accompanied by information changes, we speak of speciation. We will show in the next section that relative entropic behavior of evolving species is apparent despite the loss of resolution concerning lower level phenomena.

We may now look at a second model, one in which the level of isolation is not affected when reproductive linkages are severed. As we discussed, so long as all populations are linked by gene flow, our calculated levels of cohesion entropy fluctuate rather than show directional change toward higher entropy as we might have predicted. In this part of the discussion, we will show that entropic behavior exists nonetheless, although it may not be immediately manifest.

In fig. 4.13 we present a scenario of cohesion changes in a species composed of fourteen populations. In table 4.6 we provide the entropy values associated with each change. We assume that isolation levels for each population remain constant; furthermore, we have not forced any one population to complete allopatry. Note that the entropy levels of cohesion fluctuate between small increases and small decreases during evolution from fig. 4.13*a–d*. However, at *e*,

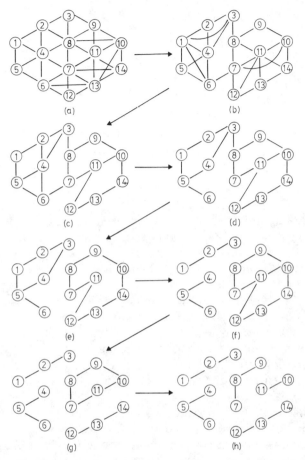

Fig. 4.13. Changing reproductive linkages in a species comprising fourteen populations.

when allopatry is achieved in part of the species (populations 1 to 6 and 7 to 14), the cohesion entropy increases dramatically (+0.095 vs. ±0.004 in previous states). Further isolation, in fig. 4.13f, produces another large increase in entropy. Following that, however, subsequent changes (fig. 4.13g–h) are characterized by entropy increases of a decreased magnitude. This indicates a trend toward minimizing entropy increases (shown graphically in fig. 4.14), which we previously showed to be associated with historical constraints (here, constant population number) on irreversible behavior.

   In the first four cases (fig. 4.13a–d), there is a relatively high probability that the species could return to its original configuration;

Table 4.6 Total Entropy of Cohesion Changes in the Hypothetical Species Shown in Fig. 4.13. (H_{cw} = cohesion entropy within subsystems; H_{ct} = total cohesion entropy; dH_c = change in H_{ct}.)

Figure	$\bar{F}\ (\sigma^2{}_F = 0)$	$-\sum p_i \log_2 p_i$	H_{cw}	H_{ct}	dH_c
4.9a	.2	3.735	—	0.747	—
4.9b	.2	3.753	—	0.751	+.004
4.9c	.2	3.773	—	0.755	+.004
4.9d	.2	3.757	—	0.751	−.004
4.9e (1–6)	.2	1.230	0.246	—	—
4.9e (7–14)	.2	3.000	0.600	—	—
4.9e (total)	.2	4.230	—	0.846	+.095
4.9f (1–3)	.2	1.500	0.300	—	—
4.9f (4–6)	.2	1.500	0.300	—	—
4.9f (7–14)	.2	2.975	0.595	—	—
4.9f (total)	.2	5.975	—	1.195	+.349
4.9g (1–3)	.2	1.500	0.300	—	—
4.9g (4–6)	.2	1.500	0.300	—	—
4.9g (7–11)	.2	2.322	0.464	—	—
4.9g (12–14)	.2	1.500	0.300	—	—
4.9g (total)	.2	6.822	—	1.364	+.169
4.9h (1–3)	.2	1.500	0.300	—	—
4.9h (4–6)	.2	1.500	0.300	—	—
4.9h (7–9)	.2	1.500	0.300	—	—
4.9h (10–11)	.2	1.000	0.200	—	—
4.9h (12–14)	.2	1.500	0.300	—	—
4.9h (total)	.2	7.000	—	1.400	+.036

that is, the changes that occurred did not substantially alter the ancestral condition. With allopatry, however, the probability of returning to the ancestral condition diminishes drastically. This does not mean that the allopatric configuration is inherently improbable, only that the ancestral condition is less probable relative to the present state of the system. Because the changes are entropic in nature, the descendant states cannot be "improbable"; as we pointed out in the simple evolution model (see also Hollinger and Zenzen 1982), entropic behavior increases the probability of the system in this sense of probability.

Why, then, does the amount of entropy increase (i.e., the rate of entropy production) diminish in this model? This can be explained easily as the result of the assumption that levels of relative isolation are not changed, producing greater gene flow among demes that remain connected. Those which remain connected through the most changes (i.e., have the longest historical association) will exhibit the greatest amount of cohesion, slowing the overall rate of entropy

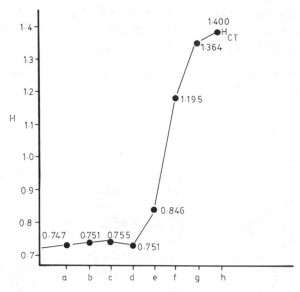

Fig. 4.14. Changes in entropy associated with changes in cohesion among populations shown in fig. 4.13.

increase. Thus, as with other cases considered in this text, the trend toward minimum entropy production in systems far from equilibrium results from historical constraints (self-organization).

Summary Remarks

We have attempted to show that reasonable estimates of cohesion and the entropy levels of cohesion can be made given the necessary data. However, estimating effective population sizes and gene flow remain problems for population geneticists and population biologists to solve. At least gene flow can be estimated grossly given certain assumptions (Slatkin 1981). Perhaps as important is that gene flow does seem to be a reality—the no-gene-flow model of Ehrlich and Raven (1969) does not seem to hold up under assumptions regarding genes at intermediate frequencies (Slatkin 1981), indicating that gene flow is present, even in low-vagility organisms. Even if it were not, however, this would not affect our thesis, since increasing isolation means increasing levels of cohesion entropy. Interestingly, the degree of fluctuation a population experiences during isolation seems to be a function of the strength of the cohesion binding populations. Our two examples indicate that small fluctuations are associated with a model that assumes the gene flow between neighboring demes is not increased with increasing isolation, while large fluctuations are associated with a model that assumes the opposite. We may speculate that this means that consolidation of information differences would be faster in cases of large cohesion fluctuations than in cases of small cohesion fluctuations. That is, the more cohesive a species before isolation, the faster its fragments will differentiate after isolation.

Species and Speciation

In at least some cases we can distinguish between species and populations by the observation that species are individuated information and cohesion systems whereas populations are not. The clearest distinction can be seen in classical biological species (see Dobzhansky 1937; Mayr 1942, 1963), in which the species is reproductively isolated from its nearest relative, that nearest relative is sympatric, and all populations within species are linked by gene flow. This distinction fails altogether when we consider asexual species in which the members have no reproductive ties and each organism is an individuated information system. Biological species share both historical continuity and cohesion. Asexual species are composed of members who share only historical continuity. Sexually reproducing species composed of entirely allopatric subpopulations (i.e., groups of demes that have gene flow between them but no gene flow with other groups of demes) represent an intermediate situation in which the subpopulations share historical continuity but cohesion is only a within-subpopulation phenomenon.

We have attempted to show how energy, cohesion, and information work in the production of entropy on the level of sexually reproducing populations. In this section, we extend our analysis to species. We will pay attention to the "simplest" case, sexually reproducing species whose populations share cohesion bonds. We do so because the relationships between cohesion and information are most clearly demonstrated in such systems. We bypass asexual species and sexual species with allopatric demes for the moment, not because they form exceptions to our theory but because they lack cohesion bonds and thus entropy production involves only information (asexual species) or information and within-population cohesion (sexual species with no gene flow between demes).

Modes of Speciation

Speciation is a process that increases the number of species in a clade or, more rarely (never?), reduces the number of species in a clade without lineage extinction. Modes of speciation that involve an increase in the number of species at any particular time may be termed *additive speciation* (Mayr 1963; Wiley 1981a), whereas the theoretical possibility of two species melding to form a third species may be termed *reductive speciation*. Species may also absorb other species, causing their extinction. This phenomenon, discussed by Harlan and deWet (1963), may be thought of as a mode of extinction since the outcome does not involve the establishment of a new species. Species may also transform through time by anagenetic processes such as those discussed previously. Although some authors recognize this phyletic evolution as a mode of speciation ("phyletic speciation"), the practice of arbitrarily subdividing an evolving lineage is not considered here as a mode of speciation (Wiley 1981a, 38–41). Previous workers have recognized a variety of different modes of additive speciation, which have been classified and discussed by Bush (1975a), Endler (1977), and Wiley (1981a), who also provide a brief history of the literature.

Allopatric Speciation. Allopatric speciation results when an ancestral species is physically separated into two or more geographical populations and at least one of these populations differentiates into a new species.

The models proposed for allopatric speciation have the common property of assuming that the cohesion of the lineage is destroyed by events extrinsic to the organisms themselves. Such events may involve geological changes such as the rise of a mountain chain or changes in the drainage patterns of a river system. They may also involve climatic changes such as periods of drought or restriction of populations to climatically suitable but disjunct regions.

Parapatric Speciation. Parapatric speciation results when at least one population of an ancestral species differentiates into a new species in spite of the fact that it maintains a narrow zone of reproductive contact with the rest of the ancestral species. The model proposed by Endler (1977) assumes differentiation along an environmental cline and a decrease in fitness of heterozygotes along the contact zone between the two populations. The result will be two species with ranges that abut (parapatric ranges) and perhaps a narrow band of sympatry.

Alloparapatric Speciation. The alloparapatric model is, loosely, a hybrid between the allopatric and parapatric models. It assumes that initial differentiation takes place in allopatry but that final differentiation takes place in sympatry, with direct genetic interactions among the daughter species reinforcing the differences between them by eliminating hybrids. Thus, isolating mechanisms (mechanisms that prevent gene exchange between sympatric species—see Dobzhansky 1951; Mayr 1963) are perfected in sympatry. This mode has been criticized by Paterson (1978). His analysis includes the rejection of the idea that species are the direct result of environmental selection for adaptation and obviously has wider implications, with which we largely agree.

Stasipatric Speciation. As outlined by White (1978), stasipatric speciation is a mode by which a chromosomal mutation leads to the establishment of a new species. The mutation is envisioned as arising in a population belonging to a species of low vagility. Individuals homozygous for the new chromosome arrangement are more fit than heterozygotes, and the population becomes fixed for the new arrangement through drift or a meiotic bias toward the new arrangement ("meiotic drive"). Dispersal from the original population results in establishing an expanding zone of parapatry with the ancestral species (thus, the mode has been termed *parapatric speciation*—see Endler 1977; White 1978).

Sympatric Speciation. Sympatric speciation may be thought of as an umbrella term for a number of modes of speciation that do not require some form of geographical isolation (partial or complete) for speciation to occur. The usual mode discussed in the literature is what may be termed (following Wiley 1981a) *ecological sympatric speciation* (Thoday and Boam 1959; Maynard Smith 1966; Bush 1975a, 1975b; Richardson 1974). It involves habitat segregation. Although this mode is controversial (see Mayr 1963; Paterson 1978), other sympatric modes are not. *Speciation via hybridization* is common in plants (Grant 1971), as is *speciation via apomixis*. Some modes may be "sympatric," depending on one's definition of sympatric. There is certainly sympatry of populations in the parapatric mode and in the stasipatric mode. Conversely, some might argue that habitat segregation is a kind of "microallopatry." However, we are not concerned with these arguments because they are largely semantic. We will distinguish "sympatric" from "allopatric" and "parapatric" by simply saying that sympatric speciation is an event requiring no geographical component, whereas allopatric and parapatric speciation have a geographical component.

Table 4.7 Four Classes of Linkage Pattern Changes When Information Changes Are Great Enough to Disrupt or Prevent Cross Linkages Between Phenotypes

	New pattern established and maintained	New pattern not established or maintained
Old pattern maintained	I. Immediate sympatric speciation	II. Stasis
Old pattern not maintained	III. Ecological sympatric speciation; parapatric speciation, allopatric speciation, or anagenesis	IV. Extinction (local or species)

Speciation and Entropy Changes

One assumption implicit in our discussion of changes within species was the assumption that changes in information were not of the magnitude that would cause the population to become completely disorganized. Complete disorganization may be manifested in two ways, extinction or lineage splitting. The most radical change that we have covered thus far is anagenesis, change without complete disorganization. A species that avoids extinction and becomes disorganized is a candidate for speciation. Speciation may result from the direct interplay of informational change and its effects on cohesion, or it may result from an extrinsic event that causes disorganization (loss of cohesion) followed by subsequent changes in information. Sympatric, parapatric, and stasipatric speciation are the modes of speciation in which cohesion and information changes are coupled. Allopatric speciation can occur with a change in cohesion followed by a change in information that is uncorrelated with the change in cohesion.

We may characterize some modes of speciation in a similar manner to changes within species by reference to pattern changes shown in table 4.2. When we translate the outcomes (table 4.7), we can see that these modes fall into either case I or case III. If cases II and IV obtain, speciation is not the result. In fig. 4.15, we present some hypothetical curves representing the changes in entropy we expect to occur during speciation. In fig. 4.15a,b, we plot changes in information for certain phenotypes (see S_x) and changes in cohesion (C). They are plotted on the same scale only for ease of comparison.

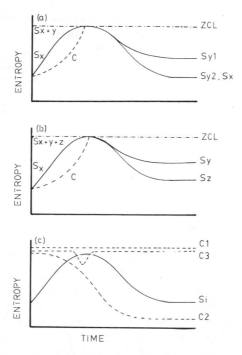

Fig. 4.15. The dynamics of speciation. a = Speciation with the persistence of the ancestral species; b = speciation with extinction of ancestral species and formation of two descendant species; c = speciation in noncohesive systems. ZCL = Threshold at which evolutionary changes among populations become irreversible; S_x = entropy state of plesiomorphic information system; S_y and S_z = entropy states of apomorphic information systems; *solid line* = information; *dashed line* = cohesion(C). (Modified from Wiley and Brooks, *Syst. Zool.* 31 [1982]:1–28.)

The line labeled "ZCL" represents what we have termed the *zero cohesion line*. Where it lies for some speciation events depends on the particular speciation event itself and the kind of information that is changing. Allopatric subpopulations are already at the ZCL; a monotypic species is far or close to the ZCL, depending on what information changes might occur in the future. Thus, ZCLs really represent a posteriori acknowledgement that evolutionary change in a species reached a point where a potentially reversible disjunction became irreversible. A "biological phase transition" occurred at that given level of complexity (Brooks and Wiley 1984). Active phase transitions are associated with parapatric and sympatric speciation; passive phase transitions are associated with allopatric speciation. Active phase transitions involve changes in information causing changes in cohesion. Passive phase transitions involve changes in cohesion (allopatry imposed from without) and "passive" informational changes whose effect or lack thereof on cohesion cannot be assessed until sympatry is achieved or until crosses resulting in sterility or lowered fecundity are measured experimentally. Of course, the longer allopatry obtains, the greater the chances that a passive phase transition will occur.

Reconsider table 4.7 and fig. 4.15. Case I is immediate, one-generation, sympatric speciation with the species reaching ZCL immediately. Such a situation would obtain in shifts from sexual to asexual reproduction. The result is two species, the ancestor (S_x, fig. 4.15a) and a descendant (S_{y1} or S_{y2}). The same would apply to speciation by hybridization and speciation as the result of autopolyplody. Case III is more complicated because the patterns are more complicated. It is not simply a matter of new and old patterns but the change from an old pattern through time, which involves interactions between the entities. Case III results include ecological sympatric speciation, parapatric speciation, and allopatric speciation. In sympatric and parapatric speciation, information changes. This would also apply to speciation via hybridization except the result would be three species, the two ancestors and the hybrid descendant. It would also apply to autopolyploid species. Case III sympatric and parapatric speciation progressively affect reproductive linkages, forcing an increase in the entropy level of cohesion until the ZCL is reached. In either case, the shape of the curves would depend on the kind of information changes that occur. We may imagine two possibilities: (a) the establishment of a new phenotype whose hybrids with the ancestral phenotype are selected against and (b) a sequential progression of phenotypes departing more and more from the ancestral phenotype as time goes on. The curves associated with possibility a result from the increase in the members of the new phenotype, with the species progressing to maximum polymorphism or polytypism and the occurrence of hybrids being progressively restricted to F_1 generation crosses. Thus the curve represents the interaction of information and cohesion in terms of changes in information. Under possibility b, there is an increase in information complexity as well as an increase in phenotype frequencies. Thus, the curves track increases in intrinsic information content of one or more phenotypes as well as the dynamics of their interactions. A measure of the intrinsic change in information complexity is the difference in entropy values between the results. Fig. 4.15a graphs changes in which the ancestral species survives a speciation event (see peripheral isolation, Wiley 1981a). There may be no net change in the entropy level of intrinsic information if the descendant species S_{y2} has substituted information. In such a case, net increase in entropy is restricted to cohesion relationships. There may be a change if the descendant species S_{y1} has added information, and a measure of this change is the distance between S_x and S_{y1} on the entropy scale. In such a case, entropy changes in the system involve both cohesion and information.

Speciation may also involve the extinction of the ancestor (Wiley 1981a). One possible outcome of this is shown in fig. 4.15b, with both descendants, S_{y2} and S_y, showing intrinsic increases in information.

Entropy changes in sympatric and parapatric speciation are the easiest to model because they involve the direct interplay of information and cohesion. Allopatric speciation differs because cohesion linkages are disrupted by extrinsic events that split the ancestral species. This allows the two (or more) allopatric populations to vary without having to integrate their information systems. The effect is a functional lowering of the ZCL, with speciation occurring as a within-population phenomenon similar, in each of the separate populations, to the curve shown in fig. 4.2. The main difference between allopatric speciation and the sympatric and parapatric modes is that in the allopatric mode information changes may lag behind cohesion changes between populations, whereas in the sympatric and parapatric modes, cohesion changes lag behind information changes because information changes are causing cohesion changes. Thus the allopatric speciation model involves the extrinsic (geographical) disorganization of a species. The cohesion of the species, between populations, is already at the ZCL. Any population experiencing an increase in information complexity of sufficient magnitude to establish a new canalized information system becomes a new species. Information changes within the affected population are labeled "S_1" in fig. 4.15c.

There are other modes of speciation involving disorganized species (table 4.7, fig. 4.15c). Asexual speciation is determined entirely by changes in information systems (fig. 4.15c, line labeled "C_1"). Speciation via hybridization involves the temporary establishment of reproductive linkages between two species (fig. 4.15c, line labeled "C_3"). The theoretical possibility of reductive speciation involves the permanent fusion of two species (fig. 4.15c, line labeled "C_2").

The dynamics of speciation, dS, is always greater than zero, indicating that speciation is an entropic phenomenon. This is summarized by the following heuristic equations in which dS is the change in entropy of a monophyletic group, and S_x, S_y, and S_z are the entropy levels for species X, Y, and Z (Wiley and Brooks 1982).

1. Persistence of the ancestral species (this includes the special case of stasipatric speciation):

$$dS = S_x + S_y - S_x > 0$$

2. Extinction of the ancestral species:

$$dS = S_y + S_z - S_x > 0$$

3. Asexual speciation:

$$dS = S_x + S_y - S_x > 0$$

4. Reduction speciation:

$$dS = S_z + \infty + \infty > 0$$

5. Speciation via hybridization:

$$dS = S_x + S_y + S_z - S_x - S_y > 0$$

Comment on Punctuated Equilibrium

Eldredge and Gould (1972) helped focus attention on speciation and the fossil record by proposing what has become known as the *punctuated equilibrium model.* Two basic observations about the fossil record form the basis of this model. First, the morphologies of various species remain relatively static for long periods of time, rather than changing gradually and continually. Second, when new morphologies appear, they are markedly different from the ancestral morphologies, and they replace the ancestral forms rather rapidly. To explain these observations, proponents of punctuated equilibrium invoked peripheral isolates allopatric speciation, in which new species originate in small populations in peripheral habitats (meaning that the likelihood of finding fossil evidence is greatly reduced) before dispersing into the ancestral range and displacing the ancestral species (meaning that the new species would appear rather suddenly and displace the old species rather rapidly). In addition, they devoted attention to concepts of developmental constraints, particularly those associated with Goldschmidt's (1940) notions of "macroevolution," in order to bolster the plausibility of finding markedly different morphologies.

Much controversy has surrounded the punctuated equilibrium model, primarily concerning claims of stasis and punctuated change versus consistent and gradual change. We believe that evidence adduced by geneticists concerning polygenic inheritance leads us to accept the notion that many characters will show gradualistic change, whereas evidence adduced by developmental "structuralists" (see chapter 3) leads us to accept the notion that some characters will show "punctuated" changes. We also expect many, perhaps even most, species to originate in geographical isolation and thus expect to find evidence of peripheral isolates allopatric speciation. Thus, depending on the developmental genetics of the characters that marked a given species' evolution and divergence, and depending on the geographical context of those changes, we expect to find evidence supporting both punctuational changes and gradualistic changes. Our theory predicts that in either case, population variance for the characters undergoing divergence should increase prior to speciation and that descendant populations will be initially low-variance assemblages. Both Williamson (1981), who presented evidence in favor of punctuated equilibrium, and Rose and Brown (1984), who presented evidence in favor of gradualism, noted an increase in population variance followed by a transition to low-variance descendant populations in their studies.

Species, Phylogenetic Tree Topologies, and Entropy

Different models of speciation may predict different phylogenetic tree topologies (Wiley 1981a). For example, peripheral isolation results in the persistence of the ancestor (fig. 4.16a), whereas a large-scale geographical subdivision may cause ancestral extinction because neither descendant can be "identified" as the ancestor (rather both "are") (fig. 4.16b). Another situation might be the budding off of two peripheral isolated descendants (Y and Z in fig. 4.16c) producing a dichotomous topology that will be interpreted as an unresolved trichotomy (fig. 4.16d). Yet another possibility is a reticulate tree resulting from hybridization (species M in fig. 4.16e). We may calculate the statistical entropy of each of these trees if we consider them to be directed graphs. In fig. 4.16a, species X has one connection to itself (since it survives the speciation event) and one

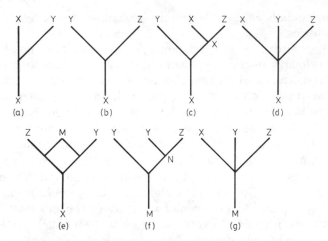

Fig. 4.16.   Seven different species-level tree topologies.

connection to Y, whereas Y has only a connection to X. The entropy of this tree, disregarding informational change, is

$$H = -(2/3 \log_2 2/3) - (1/3 \log_2 1/3)$$
$$= 0.918 \text{ bit}$$

For fig. 4.16b, X has connections with Y and Z and both Y and Z have only connections with X. The entropy is

$$H = -(2/4 \log_2 2/4) - (2)(1/4 \log_2 1/4)$$
$$= 1.5 \text{ bits}$$

For fig. 4.16c, the situation is more complicated because X is connected with itself twice and with Y and Z once for a total of four connections, whereas Y and Z are connected only to X. The total number of connections is six. The entropy is

$$H = -(4/6 \log_2 4/6) - (2)(1/6 \log_2 1/6)$$
$$= 1.251 \text{ bits}$$

For fig. 4.16d, the entropy is

$$H = -(3/5 \log_2 3/5) - (2)(1/5 \log_2 1/5)$$
$$= 1.371 \text{ bits}$$

In the hybrid phylogeny, X is connected to Z and Y, Y is connected to X and M, Z is connected to X and M, and M is connected to Y and Z. The entropy is

$$H = -(4)(1/4 \log_2 1/4)$$
$$= 2.00 \text{ bits}$$

There are some potentially interesting implications of these measures. Peripheral isolation (one of the cornerstones of the punctuated equilibrium model) emerges as the lowest entropy configuration of an outcome that results in two species. If an ancestor persists through two speciation events, the clade's entropy is lower if we consider the sequence of events to be dichotomous (fig. 4.16c) rather than a single trichotomous event (fig. 4.16d). Alternatively, speciation via hybridization is a relatively high entropy event. Interestingly, if the species of hybrid origin is intermediate in characters between the two parental species, the result on the level of character analysis will be a trichotomy (see Bremer and Wanntorp 1979; Humphries 1983; Nelson 1983; and Wagner 1983 for comments on reticulations and reconstructing phylogenies); hybrids do not always have to be intermediate and thus can appear as dichotomously related to one but not both parents. Thus, if we postulate a trichotomous phylogenetic tree, and if that trichotomy includes an ancestor, we will underestimate the entropy change if hybridization is involved and overestimate the entropy change if the descendants are peripheral isolates produced in a dichotomous fashion.

Two other types of trees are of interest in our discussion. We note that in fig. 4.16b, involving extinction of ancestor X, a higher entropy graph is shown than in fig. 4.16a. It follows that a phylogeny resulting in three species that involves ancestral extinctions should give a higher entropy value than phylogenies that do not involve extinction. Two such graphs are shown in fig. 4.16f,g. In fig. 4.16f the outcome is the same topology as in 4.16c, but the entropy value is higher:

$$H = -(2)(2/7 \log_2 2/7) - (3)(1/7 \log_2 1/7)$$
$$= 2.236 \text{ bits}$$

The entropy value in fig. 4.16g is higher than the comparable fig. 4.16d:

$$H = -(3/6 \log_2 3/6) - (3)(1/6 \log_2 1/6)$$
$$- 1.792 \text{ bits}$$

We will see that these considerations play a part in our discussions on the relationship of parsimony and evolution in chapter 5.

Since speciation involves both information and cohesion changes, and since both information and cohesion behave entropically, we should be able to integrate them. We have already seen that at the population level, cohesion constrains information changes. Previous theories of evolution have envisioned biological order emerging from the interplay of opposing forces. Our theory views speciation as a manifestation of complementary, and coupled, entropy flows.

In chapter 2, we suggested that when two complementary entropic processes are coupled, the one increasing at the slower rate will determine the behavior of the system. By analogy, the increase in information, signifying the emergence of organization, in the simple evolution model was made possible because actual entropy increases lagged behind the maximum possible entropy increases allowed by the increases in the "alphabet." Likewise, the amount of difference between maximal possible cohesion entropy and actual cohesion entropy should be directly proportional to the degree of organization of a species' breeding system.

Working in concert, entropic changes in information and cohesion should produce an additional level of organization, signified by the emergence of sexually reproducing species or populations as individuated information *and* cohesion systems. If information entropy is increasing at a faster rate than cohesion entropy, this higher level organization will be determined by the cohesion capacity. Conversely, if cohesion entropy increases faster than information, the constraint on organization of the evolving lineage will be the information capacity. These two possibilities are shown in fig. 4.17. Note that the emergent systems (H_{sp} in fig. 4.17a,b) in each case cannot be fully "diagnosed" by reference to information (a "typological species concept") or to cohesion (a "biological species concept").

To show this, the entropic status of a sexually reproducing species that comprises a number of populations with equal degrees of isolation may be estimated by the following:

$$H_{sp} = -\bar{F} \sum \frac{p_i^i p_i^c}{X} \log_2 \frac{p_i^i p_i^c}{X}$$

where H_{sp} is the entropy of the species, p_i^i refers to information probabilities and p_i^c refers to cohesion probabilities, and X refers to the total number of information microstates times the total number of cohesion microstates (required to normalize the equation; see next section). Maximum entropy would be $\log X$, and the extent to which the species is organized would be given by $\log X - H_{sp}$ (see fig. 4.17c). Hence, the capacity that changed the least would determine the species organization to the greatest extent.

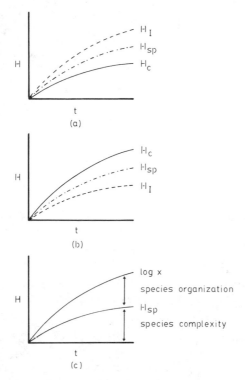

Fig. 4.17.   Schematic representations showing interaction of macroscopic (= entropic) changes in information (H_I) and cohesion (H_C) producing macroscopic changes in a species (H_{sp}). *a* = Entropic changes in cohesion less than entropic changes in information; *b* = entropic changes in information less than entropic changes in cohesion. Note that in each case the entropic behavior of the species is less than that allowed by the change in the more changeable capacity, each of which is historically constrained itself. This suggests the emergence of an additional level of constraint on entropy increases in sexual as opposed to asexual species. *c* = Relationship between complexity and organization of an evolving species; *X* = maximum possible number of information states times maximum possible number of cohesion states; H_{sp} = entropy of the species at any given time expressed in terms of observed information and cohesion configurations.

There are two components involved in the evolution of any group, namely, the topology of the phylogenetic tree (i.e., the sequence of speciation events, or changes in cohesion) and the growing amount of information potentially accessible to the evolving lineage. The tree topology includes two classes of branches, terminal and non-terminal. Nonterminal branches reflect ancestral information common to two or more sister taxa. Thus, for the simplest cases of one nonterminal and two terminal branches, the appropriate denomi-

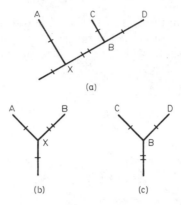

Fig. 4.18.   Phylogenetic tree (*a*) and its two component speciation events (*b* and *c*). Each slash mark denotes an apomorphic trait. (Redrawn from Wiley and Brooks, *Syst. Zool.* 32 [1983]:209–19.)

nator for establishing p_i values is 1 for each of the terminal branches and 2 for the nonterminal branch, or a total of 4. The entropy of the topology alone is

$$H = -(1/4 \log_2 1/4) - (1/4 \log_2 1/4) - (1/2 \log_2 1/2)$$
$$= 1.5 \text{ bits}$$

where the first two terms refer to the terminal branches and the third term refers to the nonterminal branch. The information accessible to the evolving lineage is distributed on the terminal and nonterminal branches of the phylogenetic tree. The character distributions can be normalized to produce a measure of statistical entropy that reflects the contributions of the topology and the characters to the entropy of the lineage simultaneously. We will illustrate this by referring to the phylogenetic trees in fig. 4.18. For fig. 4.18*b*, the probabilities for the branches are 1/4, 1/4, and 1/2, as above. In addition, there are four characters, 1/4 of them on branch A, 1/2 of them on branch B, and 1/4 of them on branch X. We then multiply the two sets of probabilities and find that

$$P_A = (1/4)(1/4) = 1/16$$
$$P_B = (1/4)(1/2) = 2/16$$
$$P_X = (1/2)(1/4) = 2/16$$

The sum of the three raw probabilities above is 5/16. Because all probabilities must sum to unity for probability measures to be comparable, we must multiply the raw probabilities by a normalizing constant. This constant is the reciprocal of the sum of the raw probabilities, in this case 16/5. The normalized probabilities are thus:

$$P_A = (1/16)(16/5) = 1/5$$
$$P_B = (2/16)(16/5) = 2/5$$
$$P_X = (2/16)(16/5) = 2/5$$

The entropy in fig. 4.18b is

$$H_b = -(1/5 \log_2 1/5) - (2)(2/5 \log_2 2/5)$$
$$= 1.521 \text{ bits}$$

For fig. 4.18c, the raw probabilities are

$$P_C = (1/4)(1/4) = 1/16$$
$$P_D = (1/4)(1/4) = 1/16$$
$$P_B = (1/2)(1/2) = 4/16$$

The normalizing constant is 16/6, and the entropy of the tree is

$$H_c = -(2)(1/6 \log_2 1/6) - (4/6 \log_2 4/6)$$
$$= 1.251 \text{ bits}$$

The entropy in 4.18c is lower than that in fig. 4.18b because more of the information is historically constrained, as synapomorphies (shared uniquely evolved traits; Hennig 1966) even though the two figures have the same topology and the same number of characters. For fig. 4.18a, the raw probabilities are

$$P_A = (1/7)(1/6) = 1/42$$
$$P_C = (1/7)(1/6) = 1/42$$
$$P_D = (2/7)(1/6) = 1/42$$
$$P_B = (2/7)(2/6) = 4/42$$
$$P_X = (2/7)(1/6) = 2/42$$

The normalizing constant is 42/9, and the normalized probabilities are

$$P_A = 1/9$$
$$P_C = 1/9$$
$$P_D = 1/9$$
$$P_B = 4/9$$
$$P_X = 2/9$$

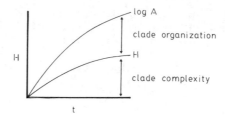

Fig. 4.19. Schematic representation of relationship between complexity and organization of evolving biological diversity. A = Maximum possible number of configurations given observed species information and cohesion systems; H = entropy of observed state of diversity for any given monophyletic group (evolving lineage).

The entropy of the cladogram is

$$H_a = -(3)(1/9 \log_2 1/9) - (4/9 \log_2 4/9) - (2/9 \log_2 2/9)$$
$$= 2.058 \text{ bits}$$

The total amount of entropy produced by the speciation events, as independent "reactions," is

$$H_{cum} = H_b + H_c = 1.521 + 1.251 = 2.772 \text{ bits}$$

However, the entropy of the lineage as an evolving system was raised only 0.537 bits by the second speciation event:

$$\Delta H = H_a - H_b = 2.058 - 1.521 = 0.537 \text{ bit}$$

Historical constraints on the evolution of species, symbolized by synapomorphies, mean that some entropy production in ancestors is retained in descendants, thus lowering the overall increase in entropy, or lowering the rate of entropy increase.

This is accompanied by an increase in the organization of the evolving clade which, because it is an emergent property of the historical constraints on the system, is *self-organization.* For fig. 4.18*b*, the organization is 0.629 bits (log 5 − 1.521) and for fig. 4.18*a*, it is 1.062 bits (log 9 − 2.058). The relationship between clade complexity and clade organization that emerges from these considerations is shown in fig. 4.19.

In fig. 4.20, we have produced a phylogenetic tree comprising twenty speciation events, beginning with a high rate of evolutionary change that slows with time. The entropy and information produced for each speciation event, the cumulative entropy and information produced, the total entropy and information of the lineage at each point, and the ΔH and ΔI values are listed in table 4.8. In this case,

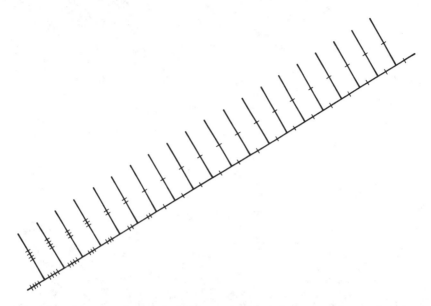

Fig. 4.20. Phylogenetic tree depicting twenty speciation events characterized by high rates of character change initially and low, stable rates later. (Redrawn from Wiley and Brooks, *Syst. Zool.* 32 [1983]:209–19.)

we make the simplifying assumption that the same amount of time elapses between speciation events. We have already discussed what happens in general to species affected by inconstant variables pertinent to speciation.

It is important to note that there is always positive entropy production; so the entropy produced by the evolving lineage is always increasing cumulatively. Note as well that the total entropy value of the lineage begins to flatten out after a number of speciation events has occurred. Examination of the ΔH values shows the lineage is approaching a non-zero asymptote of rate of entropy increase; as shown graphically in fig. 4.21. As in chapter 3, we can see that historical constraints minimize entropy increases through time and that moving from one functional level to another causes a loss of resolution of total entropy increase at lower levels. In fig. 4.22 we present a somewhat more realistic phylogenetic tree; in this case, evolutionary rates drop three times and then rise again following two of those drops (see asterisks in fig. 4.22). See table 4.9 for information about fig. 4.22 equivalent to that provided in table 4.8 for fig. 4.20. Fig. 4.23 is a graphic portrayal of the shifts in rate of entropy production. Because we have assumed constant time for

Table 4.8 Entropy and Information Values Calculated for Tree Shown in Fig. 4.20 (See note for explanation of symbols)

H_n	H_{cum}	H_{tot}	ΔH	I_n	I_{cum}	I_{tot}	ΔI
1.5	1.5	1.5	—	4.085	—	4.085	—
1.235	2.734	2.235	0.736	4.350	8.435	4.894	0.809
1.413	4.147	2.680	0.445	3.909	12.344	5.342	0.440
1.585	5.732	3.048	0.368	3.585	15.929	5.652	0.300
1.330	7.062	3.299	0.251	3.477	19.406	5.869	0.217
1.5	8.561	3.532	0.233	3.085	22.491	6.038	0.169
1.251	9.812	3.677	0.145	2.749	25.240	6.174	0.136
1.5	11.312	3.777	0.100	2.085	27.325	6.326	0.152
1.5	12.812	3.872	0.095	2.085	29.410	6.459	0.133
1.5	14.312	3.961	0.089	2.085	31.495	6.578	0.119
1.5	15.812	4.050	0.089	2.085	33.580	6.681	0.103
1.5	17.312	4.134	0.084	2.085	35.665	6.776	0.095
1.5	18.812	4.212	0.078	2.085	37.750	6.865	0.089
1.5	20.312	4.289	0.077	2.085	39.835	6.944	0.079
1.5	21.812	4.363	0.074	2.085	41.920	7.018	0.072
1.5	23.312	4.434	0.071	2.085	44.005	7.087	0.069
1.5	24.812	4.503	0.069	2.085	46.090	7.151	0.064
1.5	26.312	4.569	0.066	2.085	48.175	7.212	0.061
1.5	27.812	4.632	0.063	2.085	50.260	7.270	0.058
1.5	29.312	4.695	0.063	2.085	52.345	7.323	0.053

NOTE: Entropy and information produced by each speciation event (H_n and I_n), total entropy and information produced if speciation events were independent (H_{cum} and I_{cum}), entropy and information of the lineage at a given time (H_{tot} and I_{tot}), and change in entropy and information production (ΔH and ΔI) for cladogram shown in fig. 4.20.

speciation, the shifts are equivalent to changes in rate of entropy production. The dips in the curve represent decreases in *rate* of entropy production (dS/dt), but there is always postive entropy production ($dS > 0$) at some level. The total entropy produced is represented by the area under the curve.

Summary

Within-species evolution results from an array of intrinsic and extrinsic factors. Since the intrinsic forces are inherently irreversible, we accord them the status of ultimate mechanistic, or axiomatic, causes. These include reproduction, mutation, canalizing selection, and perhaps concerted evolution. Phenomenological descriptions of evolutionary changes brought about by these factors should be characterized by a constrained increase in information entropy similar to that predicted by the evolutionary dynamic in

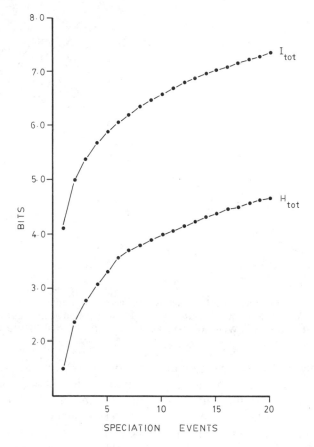

Fig. 4.21. Graph of total macroscopic entropy (H_{tot}) and total macroscopic information (I_{tot}) versus time (represented by speciation events) for phylogenetic tree shown in fig. 4.20. The cumulative amount of entropy produced is the area under the curve, whereas the curve itself represents the changes in the rate of entropy production for the lineage.

chapter 2. Many loci may exhibit equilibrium behavior in a population or species, but globally, populations' and species' information systems will always be in nonequilibrium states. For sexually reproducing species, evolutionary changes in reproductive ties will exhibit similar behavior. In addition, the interaction of information and cohesion should produce highly complex organized species.

The longer a population persists as a genealogical continuum, the more highly self-organized it will become. This historically determined emergent organization can be affected by proximal effects

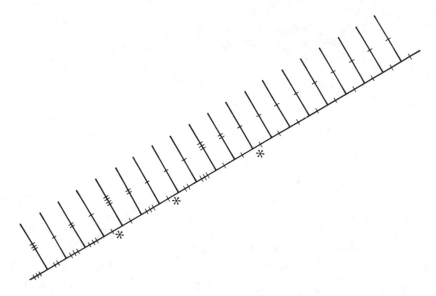

Fig. 4.22.   Phylogenetic tree depicting twenty speciation events with three episodes of reduction in rate of change (*asterisks*) and two subsequent increases in rate of change. (Redrawn from Wiley and Brooks, *Syst. Zool.* 32 [1983]:209–19.)

generally called environmental selection. However, the ability of environmental selection to impose organization should have predictable limits. To illustrate this point, we can consider four general modes of selection: those which have (1) a specific focus (i.e., a single trait or a small number of linked traits) or (2) a broad focus (i.e., the general phenotype) and those which (3) complement or (4) conflict with the emergent organizing tendencies. Modes 1 and 2 interact with modes 3 and 4 to produce four classes of selection pressures (table 4.10).

Since the essence of emergent organization is historical constraint, the aspects of a population's information and cohesion systems that can be complemented by directional selection will be predominately the historically determined ones. Phylogenetic analysis (see chapter 5) can provide direct estimates of this historical structure for a chosen species. Either narrowly (class I in table 4.10) or broadly (class II) focused selection that complements the inherent tendencies will enhance the rate at which organization appears. In such cases, the population structure will be a combination of emergent and imposed organization. To the biologist-observer, it would appear that the population had "adapted to its environment" very quickly.

Table 4.9 Entropy and Information Values Calculated for Tree Shown in Fig. 4.22 (See note for explanation of symbols)

H_n	H_{cum}	H_{tot}	ΔH	I_n	I_{cum}	I_{tot}	ΔI
1.435	1.435	1.435	—	3.565	—	3.565	—
1.404	1.839	2.144	0.709	3.181	6.746	4.248	0.683
1.435	3.274	2.621	0.477	3.565	10.311	4.488	0.540
1.061	4.335	2.887	0.266	3.261	13.572	5.061	0.273
1.378	5.713	2.900	0.013	3.207	16.779	5.685	0.624
1.556	7.269	3.485	0.585	3.029	19.808	5.621	−0.064
1.060	8.329	3.621	0.136	3.262	23.070	5.792	0.171
1.510	9.839	3.813	0.192	2.490	25.560	5.918	0.126
1.404	11.243	3.978	0.165	3.181	28.741	6.077	0.159
1.295	12.538	4.107	0.129	3.512	32.253	6.239	0.162
1.521	14.059	4.247	0.140	3.429	34.732	6.332	0.093
1.5	15.559	4.358	0.111	2.085	36.817	6.406	0.074
1.5	17.059	4.464	0.106	2.085	38.902	6.472	0.066
1.5	18.559	4.563	0.099	2.085	40.987	6.535	0.063
1.5	20.059	4.657	0.094	2.085	43.072	6.594	0.059
1.5	21.559	4.746	0.089	2.085	45.157	6.650	0.056
1.5	23.059	4.831	0.085	2.085	47.242	6.702	0.052
1.5	24.559	4.911	0.080	2.085	49.327	6.753	0.051
1.5	26.059	4.989	0.078	2.085	51.412	6.799	0.046
1.5	27.559	5.062	0.073	2.085	53.497	6.846	0.046

NOTE: Entropy and information produced by each speciation event (H_n and I_n), total entropy and information produced if speciation events were independent (H_{cum} and I_{cum}), entropy and information of the lineage at a given time (H_{tot} and I_{tot}), and change in entropy and information production (ΔH and ΔI) for cladogram shown in fig. 4.22.

Selection pressures that conflict with the inherent tendencies will be largely ineffective in cases of narrowly focused selection (class III). The more narrowly focused the selection in such cases, the less effective selection should be due to the developmental constraints on the traits. Broadly focused selection pressures that conflict with the inherent tendencies (class IV) may have a variety of effects. If such pressures are not strong enough to impose population structure contrary to the inherent tendencies in each generation, their effects should be qualitatively similar to those suggested for class III phenomena, but could persist longer. A good example of this is the sickle-cell trait. Homozygous recessive genotypes (homozygous for the sickling trait) are lethal to sublethal; thus, we would expect to find a very low rate of occurrence of the trait in any population. However, homozygous dominant genotypes are highly susceptible to malaria, whereas heterozygotes are somewhat refractory. Hence, we find that the frequency of the trait is maintained at higher levels than would be expected. The inherent trend toward reducing the

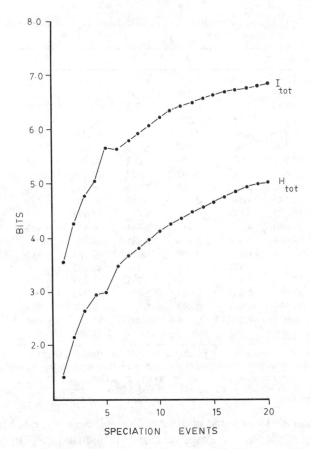

Fig. 4.23. Graph of total macroscopic entropy (H_{tot}) and total macroscopic information (I_{tot}) versus time (represented by speciation events) for phylogenetic tree shown in fig. 4.22. The cumulative amount of entropy produced is the area under the curve, whereas the curve itself represents changes in the rate of entropy production for the lineage.

frequency of the sickle-cell trait has been slowed down by the selection process.

On the other hand, if selection overrides the genealogical continuum, there is no opportunity for emergent organization to continue or to be maintained. In the short term, the frequency of occurrence of a particular phenotype or phenotype class would rise, but this imposed organization would be at the expense of the inherent organization. In the long-term, the population would remain less than optimally stable, and thus more susceptible to catastrophic occur-

Table 4.10   Four Classes of Environmental Selection Regimes

	Mode	
Selection regime	Narrow focus	Broad focus
Complementary	I	II
Conflicting	III	IV

NOTE: Regimes with a narrow focus (I) or broad focus (II) complement inherent (historical) tendencies; those with a narrow focus (III) or broad focus (IV) conflict with the inherent tendencies.

rences. The greater the discrepancy between the imposed and emergent organization tendencies, the more we would expect the emergent organization to be disrupted. On the developmental level, we would expect to see an increase in congenital defects in the population. Inbreeding programs in conflict with the inherent population tendencies will be unsuccessful in direct proportion to the degree of departure from inherent tendencies. At the population level, we would expect to see unstable population structure. For any given population, we could predict a maximum degree of difference between the inherent and imposed organization, which would constitute an instability threshold. Beyond this threshold, a population would crash, either to extinction or to the stable state it occupied historically. That is, the characteristics of a population could conceivably be altered by selection to a great degree and then spontaneously regress to states similar to those in most other populations of the species. This "regression to the species mean" or "reversion to wild type" occurs occasionally in laboratory populations of *Drosophila*.

The more effectively a selection regime imposes organization that conflicts with the historical tendencies of the population, the greater the likelihood the population will either become extinct or will revert to its historical tendencies spontaneously. Selection regimes that complement the historical tendencies can increase the rate at which a population approaches a stable state (i.e., they can slow the rate of entropy increase), but that stable state will be primarily one of historically constrained, emergent organization. These are the limits of environmental selection acting on self-organizing units. The lesson artificial breeding provides is not a model for the production of organized organic diversity, but the realization that imposed organization may be difficult to achieve (i.e., if it conflicts with the historical tendencies) and if achieved in this manner is often unstable. If initial condition constraints are important and ubiquitous, studies of natural selection (i.e., population-level studies of organism/environment interactions) should use equations in which initial condition parameters are removed. For example, Ginzburg (1980) suggested that environmental selection (which he called environmental force) be defined as a property proportional to the acceleration of population growth $[\ln \dot{N})]$. Population growth rate (r), also called the intrinsic rate of increase, is considered part of the initial conditions under Ginzburg's model and is removed from the formulations. If our theory is correct, adoption of approaches such as Ginzburg's should provide more accurate assessments of the effects of natural selection.

Speciation, the irreversible splitting of lineages, is also an entropic phenomenon. In asexual species, the pattern of speciation mirrors the pattern of changes in canalized stored information. For sexual species, the pattern of speciation will follow changes in information and cohesion. Modes of speciation in which the ancestor persists are accompanied by lesser increases in entropy than those in which the ancestor is completely transformed into descendants, except for hybrid speciation, which is accompanied by the highest entropy increase of all. We do not yet have a causal explanation for extinctions, if one exists. Perhaps there is another level of analysis, such as the "species selection" program (see Vrba 1984), which will be developed to the extent that it can provide some insight into the question of extinctions.

In considering whether or not we need a general mechanistic explanation for extinction, we must remember that there are several ways in which a species can become extinct. Perhaps the most common mode of species extinction is speciation in which the ancestor is "replaced" by its descendants. We have shown that speciation is an entropic phenomenon and that speciation involving the loss of the ancestor is more entropic than that in which the ancestor persists. It is also conceivable, although we know of no documented cases, that a species could become extinct because of developmental constraints. If developmental canalization allows progressively fewer mutants to survive, but mutations occur at a constant rate, a population could breed itself out of existence as fewer and fewer offspring survived to maturity.

We have increasing evidence of periodic mass extinctions in the past apparently due to catastrophic extrinsic perturbations (Raup and Sepkoski 1984). Such massive perturbations would decrease the number of species and thus the number of different energy-uptake pathways, slowing the increase in entropy and information resulting from evolution in all lineages. Since this is a boundary conditions effect, such mass extinctions would not necessarily result in a compensatory replacement. That is, we would not expect the same species to re-evolve after the environmental conditions returned to the precatastrophe state. Nor would we expect the same "niches" or "adaptive zones" to be filled. Rather, we would expect the surviving lineages to continue to evolve and perhaps show an increased rate of evolutionary diversification in some of those lineages *allowed by, but not caused by,* environmental opportunities made available by the extinctions. It is also possible that one species could cause the extinction of another by competition. The decrease in one species' population size would be compensated by an increase in the other species' population size. This is an initial conditions effect followed by a compensatory result.

Speciation and extinction, like birth and death, represent fluctuations that cause partial entropies associated with boundary conditions to increase or decrease and that cause partial entropies associated with initial conditions to increase at varying rates. Fluctuations that decrease boundary conditions entropies need not necessarily be compensated, but if they are, the compensation will take the form of increasing rates of entropy production in initial conditions effects.

While the entropic nature of biological evolution at this functional level is less obvious than with developmental phenomena, it is nonetheless detectable. The principles of irreversibility, individuality, and intrinsic constraints are clearly involved with evolutionary dynamics at this level. That the principle of compensatory changes also operates in this context is not quite so clear, although we predict that it does. We are struck by the similarities in relative degree of variation, as documented by electrophoretic studies, among diverse groups of organisms. Furthermore, we are intrigued by the apparent degree to which we can predict that characters that are highly variable within a population will be highly variable between populations as well (Kluge and Kerfoot 1973).

We have stressed that our theory extends rather than conflicts with neo-Darwinism. To show this, we may demonstrate that new questions raised by our theory presuppose previous empirical studies and concepts. For example, in asking "What are the *limits* of selection?" we presuppose studies of the *units* of selection. Studies of the units of selection, encouraged by neo-Darwinism, set the stage for questions about the relative role of selection in evolution. Moreover, now that we have discussed developmental, population, and phylogenetic expectations stemming from our theory, it is possible to consider various concepts of fitness.

Nonequilibrium systems irreversibly approach the most stable states possible in a given environment, and this is characterized by minimum rates of entropy production. Systems whose macroscopic behavior is constrained by their initial conditions exhibit increases in macroscopic information complementary to their entropy production. Concepts of fitness generally involve some notion of functional efficiency. The tendency to assume spontaneously the "fittest" possible configuration will be trivially true for biological systems because of the physical laws they obey, not because of any peculiarly biological factors. For example, organisms should become more efficient as they develop, achieving maximum "fitness" as reproducing adults. Prigogine and Wiame (1947) showed that this could be predicted from elementary postulates of nonequilibrium thermodynamics. Gatlin's concept of optimal coding as fitness would be included in this category. Thus, this inherent sense of fitness in biological systems is directly related to the causal laws affecting them, but it does not add anything to our understanding of their evolution beyond that provided by the basic macroscopic physical model.

There is another concept of fitness that is more restricted to biology. The members of any population can be viewed as a collection of developmental programs (as simulated in the computer model), each specifying a different minimum rate of entropy production that can be achieved during ontogeny. Given different environmental conditions, some of those developmental programs may produce more offspring than others and will tend to predominate in the population. Most evolutionary biologists consider those genotypes leaving the most offspring to be the most fit; there is nothing trivially true about this. However, there is nothing inherently irreversible about this sense of fitness. Since eliminating some evolving developmental programs constrains increases in population entropy due to mutations, this environmental influence would tend to limit the rate of entropy production, but only if the population were exhibiting irreversible behavior caused by other factors. Therefore, this aspect of fitness does not relate to any inherently macroscopic phenomenon, despite its usefulness in characterizing the relatively more and less successful genotypes in populations during particular time periods and under particular sets of environmental regimes.

Finally, what about a sense in which some species of some clades are "more fit" than others? Since evolution within species and evolution of new species are historically constrained entropic processes, we could define species or clade "fitness" in the same way we defined it for a developing organism. And, for the reasons discussed above, it would be trivially true that all species and all clades would tend to assume states of "maximum fitness" while they persisted. Thus, we expect evolutionary processes to embody a causal but trivial and a nontrivial but noncausal sense of fitness, neither of which affords us additional insight into the evolutionary process.

To us, the most startling and most hopeful outcome of this chapter is the recognition that the empirical core of neo-Darwinism, namely population biology, can be accommodated within our theory. This is startling because, until now, the success of population biology has been taken by some to indicate de facto the superiority of neo-Darwinism. But if population biological phenomena can fit equally well within two different theories, the criteria for choosing between the theories will not be population biological ones. This is a hopeful finding because it gives us a clue that we are not dealing with two opposing but two complementary theories, one of which is more general than the other.

· 5 ·

Mapping Historical Change

Having predicted that a hierarchy of species should emerge from evolutionary dynamics, it is incumbent upon us to propose a method for discovering the actual hierarchy that did emerge, or at least as many parts of it as we can. This is a question in systematic biology.

This chapter is concerned with two related questions. First, should biological classifications play an active or a passive role in analyzing evolutionary theory? And second, *can* biological classification play an active role in such analyses? We assume that no evolutionary biologist would wish to rule out, a priori, any possibly useful tool for studying biological evolution. However, biological classifications are, at the very least, descriptive constructs, and some might think that there should be a dichotomy between description and explanation in evolutionary biology. Such a view is not widely considered to be philosophically valid or likely to lead to productive empirical studies (Harré 1972; Churchland 1979). We will assume that there really is no question as to whether or not classifications should play an active role in evolutionary biology. The real question, then, is whether or not they *can* play an active role.

It seems to us that there are two current perspectives in biology concerning classifications. The first is traditional. Biological classifications are devices for summarizing the organic diversity with which we are faced and for which we are trying to find explanations. Eighteenth-century natural history featured a taxonomy-oriented, static view of nature. This has led some to believe that taxonomy is inherently static. Within an evolutionary framework, classifications are the repositories of information not only about the attributes of organisms but also about the evolution of those organisms gleaned from other areas of research. In this view, biological classifications play a passive role in evolutionary biology, serving only as pillars of stability, guideposts to our current state of (theoretical) knowledge. This is analogous to what Churchland (1982) characterized as the "functionalist" or "top-down" approach to studying the evolution of human cognitive activities.

Functionalist approaches to classification produce a systematics with a passive role in evolutionary biology. A top-down approach to classification requires that all categorical levels be recognized beforehand, then classified, and evolutionary explanations provided for those categories defined a priori. For example, one might ask how to classify fishes so as to represent their evolution, without asking if all "fishes" are each others' closest relatives. Because a priori designation will be based on nonsystematic criteria, the classificatory process is passive with respect to evolutionary theory. This approach rules out any possibility that novel relationships may emerge from systematic analyses per se. Thus, it can never lead to the elimination of commonsense groupings that are not products of evolution, such as "reptiles," "dicotyledonous plants," "fish," or "algae." Indeed, it can even lead to the formulation of evolutionary explanations for such nonexistent groups.

In the past fifteen years or so, there has emerged a vocal group of systematists, following the lead of Hennig (1966), who have made the extraordinary claim that systematic analysis can and should be an empirical tool for analysis of evolutionary theory. A case has been made for the use of and reliance on systematic techniques in biogeography (Croizat, Nelson, and Rosen 1974; Nelson and Platnick 1981), speciation models (Wiley 1981a), coevolution (Brooks 1979, 1981b), evolutionary ecology (Brooks 1980, 1985), and functional morphology (Lauder 1981). Such an approach is at odds with the traditional, functionalist view. Churchland (1982) contrasted functionalist approaches with what he termed *structuralist* or *bottom-up* approaches. This new group of systematists is clearly espousing a structuralist approach to systematics. This would be an active process of using basic observations to reconstruct the pattern (i.e., "structure") of natural diversity. From those patterns, the validity of hypotheses about evolutionary mechanisms may be scrutinized. For example, no explanation about the "evolution of reptiles" can be valid if a structuralist approach discovers that "reptiles" are not each others' closest relatives (i.e., that some "reptiles" are more closely related to nonreptiles). More generally, if a structuralist systematic approach fails to find empirical support for such grade groups, and one evolutionary theory requires grade groups to be real evolutionary entities, systematic techniques would have been successful, by themselves, in showing that this theory had no empirical support for one of its major premises. In this way, systematic techniques would become active tools for testing evolutionary theory.

Of course, no structuralist approach is theory-free, and any such approach must be justified by reference to the causal theory from which it is derived. In addition, if there is more than one causal theory of evolution, it is possible that more than one structuralist approach to systematics can be formulated. Under those conditions, studies of the differences between systematic techniques may become areas of critical research into the relative merits of the theories. Our basic position is this: (1) biological classifications should be concerned with evolutionary theory, if for no other reason than no part of science can ever be free of theory; (2) biological classifications should play an active role in evolutionary theory; and (3) classifications can play such a role if they are derived from an explicitly structuralist systematic technique. This chapter is devoted to the elucidation of such a technique deduced from the principles of our theory and comparison of it with some current systematic techniques.

Linguistic and Structuralist Approaches to Systematics

The goal of a structuralist approach in linguistics is to discern basic properties and laws pertaining to a system and its components, construct a rational taxonomy (also called a syntax) based on those laws and properties, and then (hopefully) read off the functional attributes of a linguistic system as analytic statements from the taxonony thus derived. Gatlin (1972) suggested that any set of symbols ordered according to a set of constraints represented a message in a language defined by the set of constraints:

> A linear sequence of symbols is the indispensable vehicle of communication between higher organisms. Of course, two- or three-dimensional objects can convey information and in some cases are superior to language; but they are limited in a way that the one-dimensional vechicle is not. Any two- or three-dimensional object and, in fact, any general situation can be described by language and thus reduced to the one-dimensional case.

Biological classifications are linear representations of information about characteristics of organisms arranged in a hierarchical manner according to some set of rules (Brooks 1981a); therefore, different classificatory methods can be viewed as different languages (or idiolects of the same language, as we will come to see). Earlier, Platnick and Cameron (1977) investigated methodological similarities between phylogenetic systematics and methods for comparing different human languages or for performing textual analysis. Since the linguistic analogy seems to be not only useful, but also highly appropriate, we will develop our argument along those lines.

We will assume that the purpose of a biological systematic technique is to arrange a number of species in a particular manner. We will also assume that a systematic technique compatible with our theory will arrange a group of species in the manner in which they arose during the course of evolutionary descent. The only way such a technique can do so is if information changes are orderly enough to permit such analysis. This would be analogous to being able to arrange a number of words to produce a meaningful sentence. Thus, a systematic technique is analogous to a linguistic *generative grammar*. In linguistic terms, we will develop a model generative grammar corresponding to our theory of the process of evolution. Throughout this chapter, we will use certain linguistic terms coopted primarily from Chomsky (1965) and references therein.

Consider species to be analogous to words. A sentence is made from words according to rules embodied in the generative grammar of the language in question. The history of a clade (natural group, monophyletic taxon) may be reconstructed by ordering species according to rules of a systematic technique applicable to studying evolution. Consider characters to "spell" species and groups of species, that is, to characterize and individuate them. Finally, consider phylogenetic trees to be analogous to sentences. A tree is composed of species whose order is determined by the characters they exhibit. We may determine this order by applying the rules embodied in our systematic technique.

Just as a sentence has a syntax, a particular ordering of words, a tree has a syntax, a particular ordering of species. In both cases, the particular ordering implies a *surface structure* of the sentence. In linguistics, surface structure refers to the descriptive adequacy of the syntax (Chomsky 1965). In particular, it is concerned with the manner in which words are pronounced (and spelled if the language is written) and the correct order of words. If we understand the surface structure of a language, we can pronounce the words and make syntactically correct sentences from these words. In systematics, this correct surface structure refers to an adequate (i.e., minimally meaningful) description of the degree to which a tree is recognizable as being a product of that particular generative grammar. Ergo, the description of the species and the order in which these species appear on the tree is the degree to which things are ordered according to the mechanical rules of syntax for that generative grammar. Of course, the analogy is not exact. Most words of a language are class constructs and thus can be used over and over again, whereas the characteristics and names of species concern individual entities. Nevertheless, the rules of syntax are analogous.

There is also a deep structure for any given syntax. This refers to any underlying regularities in the syntax that transcend particular cases. Deep structure refers to the explanatory adequacy of a generative grammar (Chomsky 1965). In this case, the deep structure is derived directly from a causal theory of biological evolution. It is at this point that we note a difference between our analogy of systematic techniques as generative grammars and real languages. In written and spoken language, the exact meaning, or *semantic content,* of messages is defined by "rules" external to the syntax, that is, the understanding between writer and reader or speaker and listener. For systematic generative grammars, the "rules" are inherent in the properties producing the syntax. Thus, we expect the deep structure to impart semantic content to the message directly.

We may conceive of several different "languages" of systematics, each with its own syntax and surface structure. For example, phenetics is a systematic language with a syntax that orders species on a branching diagram according to various estimates of "overall similarity" (Sneath and Sokal 1973), including both primitive and derived characters. In contrast, phylogenetic systematics is a systematic language that orders species on a tree according to whether or not they share derived characters, or synapomorphies (Hennig 1966). In either case, if the species are ordered correctly according to the syntactical rules applied, the sentence has the correct surface syntax. For our purposes, the preferred systematic technique would be one with a deep structure based on genealogical descent. Further, we seek a technique based on our particular theory of genealogical descent with modification. Such a technique would allow us to discover the surface structure of life's diversity. We might then compare various available techniques to see if one or more of them is the same technique we have generated from our causal theory. Finally, we may use such a technique to investigate the deep structure of particular portions of the surface structure we have generated. The

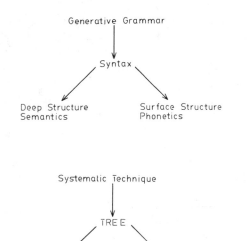

Fig. 5.1. Relationships among major components of formal language analysis (*top*) and formal systematic analysis (*bottom*). (Modified from Brooks and Wiley 1985.)

logical relationships among the various linguistic components discussed above are summarized in fig. 5.1.

Syntax

The central idea behind a successful systematic technique is that we can use it to determine the surface and deep structure of a given array of species by repeated application of a set of formal operations. These formal operations are performed on objects of an elementary class of objects or entities, such as the characteristics of individual organisms, populations, or species. Successful elucidation of surface structure leads to an interpretation of deep structure. The result is a descriptive language of evolution.

Any language is a subset of all possible combinations of a finite set of symbols (for evolution, the potential number of symbols is infinite, but at any given time we are dealing with a finite set). We may consider the "words" that make up the language of evolution to be populations or species that form lineages through time. The "alphabet" comprising these words is the characteristics of the individuals belonging to the lineages (and, of course, lineages are dynamic and their characteristics can change through time). If we wish to write a sentence in a language (be it evolution or French), we must have two things: (1) a starting point and (2) a set of rules by which we can arrange our words so the sentence makes sense. "Makes sense" refers to the context of the sentence, both its surface structure and its deep structure. What we wish to do in this section is develop the starting point and the rules for arranging species on trees. Our goal is to produce a generative grammar of evolution. We begin in a formal manner, setting out our definitions and goals symbolically, following Papentin (1980):

Let L = the language (evolution)
  w = a word (taxon)
  G = the generative grammar (the systematic technique)
  c = a subset of
  $\in$ = an element of
  $\cup$ = union of
  $\cap$ = intersection of

Further, let $G: = [A_T, A_N, S_o, R_n]$

where A_T = terminal alphabet, A_N = nonterminal alphabet, A = joint alphabet, S_o = starting point or start symbol, and R_n = set of "n" rules.

We shall stipulate some conditions:

$$A_T \neq \emptyset$$
$$A_N \neq \emptyset$$
$$A_T \cup A_N = \emptyset$$
$$A_T \cap A_N = A$$
$$S_o \in A_N$$
$$R_n = [(u_1,v_1),(u_2,v_2),\ldots,(u_n,v_n)]$$

where $(u_1,v_1),(u_2,v_2),\ldots,(u_n,v_n)$ is a set of rules for organizing information bits into a message.

Finally, let

$$L \subseteq A_T$$
$$w \in L$$

We may now interpret some of the symbols. If L is the language of evolutionary history and $w_1,w_2,\ldots,w_n$ represents the species produced during evolutionary descent, then

A_T = characters for each taxon
A_N = range of possible transformations among the characters
A = transformation series for all the characters
S_o = rooting point of the phylogenetic tree
R_n = set of rules for organizing each character into a meaningful part of the message
(u_n, v_n) = rules used to organize each unit of formal operation

We will assume that any biological theory of systematics will use the same empirical base, which is a group of characters (A_T) for a particular group of species (w). We may proceed to develop our method considering the syntactical structure defined by S_o and R_n.

The general syntax. The syntax of our theory is a tree of species hierarchically arranged in such a way that each species is placed at its genealogical point of origin. A systematic technique based on our theory must be capable of achieving this goal.

The rooting point (S_o). Our theory specifies that further modification of an information system is constrained by the intrinsic properties of that information system. The rooting point for a particular statement about an array of species must therefore be the information system of the ancestor of that array of species. The problem with rooting by the ancestral information program is that we rarely, if ever, have access to the ancestor in question. An alternative is to estimate which character in a transformation series is likely to be the character specified in the ancestral information program. Such an estimate can be made by using an "outgroup", which is defined as a species or group of species closely related to the group under analysis without being a member of that group. The utility of an outgroup for determining relevant parts of the ancestral information program has been described extensively (see Wiley 1981a; Lundberg 1972; Watrous and Wheeler 1981; Maddison, Donoghue, and Maddison 1984).Since outgroups can also evolve, more than one outgroup may be required. This approach is in accord with the "bottom-up" methodology of structuralist programs in general.

Another line of thought is directly deducible from our theory and leads to the same rationale for rooting. We have asserted that phylogenesis as a dynamic process tends toward minimizing entropy increase. So long as speciation is a process characterized by the consolidation of ancestral potential information into descendant stored information, such a tendency will occur. One measure of the utility of using an outgroup as a start symbol is to determine if such a symbol leads to a minimum entropy description of the characters that have evolved. Further, it should also lead to sentences that have the most meaning in terms of deep structure, or relevance to causal theory. In short, our theory predicts that outgroup rooting will result in minimum entropy surface structure while at the same time yielding the most informative deep structure.

Ontogenetic criteria such as the application of "von Baer's law" provide valuable auxiliary information in determining which of a homologous pair, or series of homologues, is representative of the ancestral information system because they are also expressions of the principles of historical constraint and minimum entropy production. It appears that such criteria are not independent of the outgroup criterion (Kopaska-Merkel and Haack 1982; Watrous 1982; Kluge 1985; Brooks and Wiley 1985;). However, there is a strong relationship between ontogeny and phylogeny that may be used in any systematic technique based on our theory. We must not forget, however, that ontogenetic principles such as von Baer's law are subject to exceptions analogous to problems encountered in homology arguments.

The criterion "common equals primitive" cannot be used as a starting point in our theory. This criterion states that the member of a homologous pair of characters that is most common in a clade is probably the most plesiomorphic (see Estabrook 1972, 1978; Crisci and Stuessy 1980). The empirical reasons for rejecting this criterion have been discussed by Watrous and Wheeler (1981), who cited appropriate references. The reason such a criterion will not provide a rooting point under our theory is that it places a nonhistorical constraint on the kind of information that changes and the relative timing of that change rather than constraining the way in which any information might change if it changes at all.

Rules (R_n). The rules we derive are based on two aspects of the principle of intrinsic constraints. First, character evolution (information entropy production) is constrained by the ancestral information system; thus, not all changes are equally probable, and the same character will not evolve from two or more different ancestral conditions. Second, the extent to which speciation is a nonequilibrium phenomenon involving changes in information and cohesion is the extent to which phylogenetic ordering will produce a minimum entropy configuration of information involved in speciation.

From the first principle we may deduce that we may order species on a tree by ordering the characters these species possess. The second principle will allow us to resolve character conflicts where

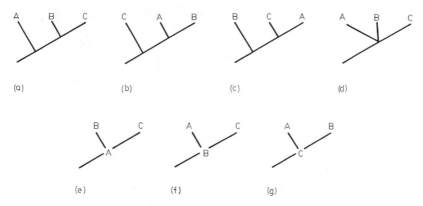

Fig. 5.2.   Seven possible phylogenetic trees given three species.

they exist because the best phylogenetic solution for any problem, given that our theory is true, should be the solution that produces the minimum entropy configuration of information changes that occurred during evolutionary descent. That is, following the dynamic proposed in chapter 2, we expect an increasing divergence from equiprobability of information as entropy increases are minimized. For example, if we have three species, the number of possible trees is seven (Wiley 1981*a*) (fig. 5.2). If we make the simplifying assumption that an ancestor is as likely to be in the analysis as two descendants, then all the trees are equally probable. Departures from equiprobability occur when characters are added to the analysis. For example, a character present in species B and C makes tree *a* in fig. 5.2 more likely than the other trees. Thus, we have departure from equiprobability. Now, if characters were randomly distributed, adding more characters would not produce a constant divergence from equiprobability. Rather, addition of characters would produce a convergence toward equiprobability. In other words, no solution would be any more likely to be true than any other solution. And conversely, in our theory, a constant divergence from equiprobability is predicted because of the nonequilibrium nature of evolution. Once a character has evolved and has been integrated into a species' information system, its future occurrence in descendants of that species is more probable than a prediction of random occurrence. Further, the very presence of such information affects future evolution of covarying characters and of the character itself. In addition, a constant or minimally variable divergence from equiprobability can be predicted. Finally, for any set of taxa, there can be only one

Table 5.1 Data Matrix for Three Taxa (A–C) and Three Characters (1–3)

Character Series	Taxa		
	A	B	C
1	0	+	+
2	0	+	+
3	0	+	+

NOTE: Zero and plus refer to different characters in each series.

historical solution. A constant divergence from equiprobability would be predicted if our analysis was proceeding correctly, that is, if each character added to the problem favored only a single tree among all possible trees. The result would be that one tree would depart from equiprobability in a manner that is completely correlated with the number of characters examined. Consider the following matrix of characters (table 5.1); the only solution is tree *a* in fig. 5.2. Now, consider another matrix of characters (table 5.2). We have a departure from equiprobability, but the departure is not constant since two trees have support. But if we use our rooting point, we may be able to show a constant departure from equiprobability (table 5.3).

Table 5.2 Data Matrix for Three Taxa (A–C) and Three Characters (1–3)

Character Series	Taxa		
	A	B	C
1	+	+	0
2	0	+	+
3	0	+	+

NOTE: Zero and plus refer to characters in each series.

Table 5.3 Data Matrix for Four Taxa (Outgroup and A–C) and Four Characters (1–4)

Character Series	Outgroup 1	Taxa		
		A	B	C
1	+	+	+	0
2	0	0	+	+
3	0	0	+	+
4	0	+	+	+

NOTE: Zero and plus refer to characters in each series, series 4 is added to show the monophyly of ABC.

Table 5.4 Data Matrix for Four Taxa (Outgroup and A–C) and Four Characters (1–4)

Character Series	Outgroup 2	Taxa		
		A	B	C
1	0	+	+	0
2	0	0	+	+
3	0	0	+	+
4	0	+	+	+

NOTE: Zero and plus refer to characters in each series.

Of course, things do not always work that way. Consider the example shown in table 5.4. In this case, we can minimize the variation in our constant divergence from equiprobability only by concluding that our data contain homoplasy, changing the matrix to that shown in table 5.5. We conclude that the original character we coded as "1" is actually two characters, "1a" and "1b." What prevents us from concluding that characters 2 and 3 are homoplasious rather than character 1? If we accept the principle that we should maximize divergence from equiprobability of the characters used to build the tree, then we accept the principle that we must consider the characters found in two species to be homologous unless there is compelling reason to consider them nonhomologous. In this case, there is no compelling reason to consider characters 2+ and 3+, found in species B and C, to be anything but homologues, respectively. Of course, we are assuming that intrinsic qualities, such as those described by Remane (see Wiley 1981a) have been fulfilled, that is, these characters have passed various tests on developmental and morphological grounds. We are not comparing apples and oranges or bird wings and butterfly wings. Thus, the way to estimate changes in information programs under the theory that evolution is a nonequi-

Table 5.5 Data Matrix for Four Taxa (Outgroup and A–C) and Four Characters (1–4)

Character Series	Outgroup 2	Taxa		
		A	B	C
1a	0	+	0	0
1b	0	0	+	0
2	0	0	+	+
3	0	0	+	+
4	0	+	+	+

NOTE: Zero and plus refer to characters in each series.

librium process is to minimize the number of homoplasy statements while maximizing the number of statements of homology. Such a system would minimize departures from a constant degree of divergence from equiprobability of possible trees. At the same time, it would maximize the divergence from independence of the characters used to construct the tree.

The simulation model presented in chapter 4 showed that splitting an ancestral species into descendants effectively converts some ancestral potential information into stored information in the descendants. For any given set of character transformations and a set of taxa, we suggest that the phylogenetic tree corresponding to the expectations of our theory is the pattern of relationships among the taxa that best reflects the stored information in the characters. Therefore, we would choose the phylogenetic tree having the highest D_1 value for the data set where

$$D_1 = \log_2 A - H_1^p$$

A = the number of plesiomorphic and apomorphic characters in the data set, and $H_1^p = -\sum p_i \log_2 p_i$, where p_i = the number of taxa united by the ith character divided by the sum of the taxa united by all characters. Only in the case of evolutionary reversals will plesiomorphic characters unite any taxa in a particular study. Because A is a constant for each data set, the value of D_1 will be determined by H_1^p, and minimizing entropy will maximize information. This provides a link between minimum entropy increases and the conversion of ancestral information into descendant information. Rule 1 may now be stated:

$$\text{Rule } 1 = [H_1^p = \min; D_1 = \max]$$

There is an alternative way of expressing rule 1, using the redundancy concept introduced in chapter 4. Gatlin (1972) wrote of two forms of redundancy, one indicating how unambiguous the sequence of individual symbols was (R redundancy, or R) and the other indicating how unambiguous the entire message was (S redundancy, or S). We will first illustrate these concepts intuitively with a simple example and then quantify them.[1]

We will consider three species and an outgroup that provides the starting point. The data matrix is shown in table 5.6. We consider species to be produced by the conversion of ancestral information

---

1. Previous attempts to relate Gatlin's concepts to systematics (Brooks 1981; Brooks and Wiley 1984) were heuristically but not computationally consistent with her views. We believe this treatment corrects those faults.

Table 5.6 Data Matrix for Four Taxa (Outgroup and A–C) and Four Characters (1–4)

Character Series	Outgroup	Taxa		
		A	B	C
1	1'	1	1	1
2	2'	2'	2	2
3	3'	3'	3	3
4	4'	4	4'	4'
5	5'	5'	5'	5
6	6'	6'	6	6'

NOTE: Prime and nonprime numbers refer to characters in each series.

into descendant information. In this example, there are ten characters that belong to that class. The least ambiguous arrangement of those characters on a tree is one in which each character represents the maximum number of repetitions of that trait among taxa. This increases the numerator of the probability associated with each character according to the formula above used to calculate the entropy. The tree shown in fig. 5.3a fulfills this criterion, whereas the tree in fig. 5.3b does not, because all the repetitions are symbolized by six characters in the former and by eight in the latter. Figure 5.3a is also less ambiguous as a whole than is fig. 5.3b, because the same trait appears as two different characters twice in fig. 5.3b. Figure 5.3a is thus preferable to fig. 5.3b by both the R-redundancy and the S-redundancy criteria.

Following the notation from chapter 4, we define R redundancy as

$$R = 1 - \frac{H_1^p}{\log_2 A_1}$$

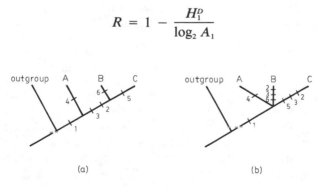

(a)   (b)

Fig. 5.3.   Two phylogenetic trees summarizing characters shown in table 5.6. a fulfills optimization criteria, whereas b does not.

or as

$$R = \frac{D_1}{\log_2 A_1}$$

Because A_1 is constant for any systematic study, increasing D_1 will increase the value of R. Increasing D_1 is achieved by decreasing H_1^p, so R and H_1^p are inversely related. Therefore, increasing entropy is associated with decreasing redundancy, as suggested in chapter 4. This provides a direct empirical and analytical link between population genetics and phylogeny, which we suggested in chapter 1 has been missing from evolutionary theory.

Gatlin's other form of redundancy, S redundancy, is defined as

$$S = I_d^{max} - I_d$$

where I_d means information density, defined as $I_d = \Sigma D_n$, and $I_d^{max} = \log_2 A_n$, where n refers to the number of characters being considered. For convenience, we will define S as

$$S \approx nH_1^p - \frac{D_n}{nH_1^p}$$

where n is the length (i.e., the total number of characters displayed on a tree) of a phylogenetic tree and $D_n = nH_1^p - H_n^p$, as per the simulation model presented in chapter 4, where $H_n^p = \log_2$ (the number of alternate trees of n length for a given set of data). For any particular case, increasing D_1 will decrease D_n, thus increasing S. Rule 2 may now be stated:

$$\text{Rule } 2 = [R = \text{max}; S = \text{max}]$$

We may derive a third rule and relate it to both rule 1 and rule 2. This rule may be termed the *synapomorphy rule* and is stated as rule 3:

$$\text{Rule } 3 = [(u_1,v_1),(u_2,v_2),\ldots,(u_n,v_n)]$$

where u_n = ignore ancestral character of a homologous series and v_n = group species (words) by the descendent characters of a homologous series.

This rule is related to rule 1 in the following manner: If we allow grouping on the basis of both primitive and derived characters, then we will not maintain $D_1 = \text{max}$, but will have D_1 varying in different directions depending on the data sets analyzed. Rule 3 is related to rule 2 in the following manner: R_{max} requires a minimum description of characters. For homologues, this means each character should be used only once since it originated only once; for homoplasies, each character should be used only the number of times it has evolved. In a transformation series of homologues, only the descendant (derived or apomorphic) characters are used because the ancestral (primitive or plesiomorphic) character has already been used at a higher level of universality than the problem we are attempting to solve (the higher level of universality is symbolized by the outgroup). For example, within Lepidosauria (snakes and lizards), the ancestral condition is the presence of forelimbs and hindlimbs, whereas the descendant condition for snakes is the absence of such limbs or their severe reduction. The presence of limbs cannot be used because it has already been used to group Tetrapoda, a larger clade that includes lepidosaurs, mammals, birds, crocodiles, and turtles. To use this character to group nonsnakes (or nonsnakes plus nonlegless lizards) would cause a violation of the R_{max} part of rule 2.

Systematic Technique Derived from Entropy Considerations

The systematic technique deducible from our theory can be stated simply. First, for any group under analysis, pick a starting point that provides the best estimate of the ancestral information system for that group. Second, estimate the transformations of the ancestral information system that have occurred in each taxon under analysis. Third, group species by shared occurrence of descendant characters. And fourth, where conflicts occur, pick the solution that maintains the closest value for constant divergence from equiprobability based on the data at hand. In terms of a generative grammar

$$G: = [A_T, A_N, S_o, R_1, R_3]$$

or

$$G: = [A_T, A_N, S_o, R_2, R_3]$$

We may now examine other systematic techniques. We will first deal with two techniques compatible with the technique developed above. Then we will deal with other approaches that are not compatible.

Phylogenetic Systematics

Phylogenetic systematics was developed by Hennig (1950a, 1966) as an explicit syntax for constructing phylogenetic trees. Hennig

(1966) stressed the primacy of genealogical relationships, the hierarchical structure of such relationships, and the fact that only shared descendant characters (termed *synapomorphies*) could be used to construct genealogies (i.e., phylogenetic trees). Specific techniques for employing Hennig's syntax as well as different views as to the nature and philosophy of Hennig's methods are now available in book form (Eldredge and Cracraft 1980; Nelson and Platnick 1981; Wiley 1981a) and have been widely discussed in the journal *Systematic Zoology*. Basically, Hennig's syntax may be summarized thusly:

1. For any group under analysis, pick the genealogically closest relatives (the outgroup or outgroups) for comparison.

2. Estimate the evolutionary transformations of any set of homologous characters in the following manner: Of each character series, that character which occurs in the outgroup and in at least one member of the group under analysis is the ancestral (= plesiomorphic) character.

3. Test each possible tree of genealogical relationships by the distribution of each shared derived character (synapomorphy) and reject all trees that are not congruent with this character distribution.

4. Where conflict occurs, that is, more than one tree is supported by synapomorphies, pick the tree that maximizes the number of possible synapomorphies.

We can see that Hennig's (1966) phylogenetics is an exact parallel to the syntax developed from nonequilibrium evolution. Steps 1 to 3 need no explanation to show their equivalence. Step 4, maximizing the number of synapomorphies, ensures that the divergence from equiprobability will be as constant as possible. In the case of data sets with much homoplasy, we will attain a lower value of divergence from equiprobability than in cases where there is little homoplasy. Phylogenetic systematics may be symbolized as a generative grammar in the following manner:

$$G: = [A_T, A_N, S_o, R_1, R_3]$$

The Wagner Algorithm

The Wagner algorithm, developed from Wagner (1961) by Kluge and Farris (1969; see also Farris 1970; Farris, Kluge, and Eckardt 1970a, 1970b), is another systematic technique compatible with our theory. It differs from phylogenetic syntax in the following manner. Rather than building a tree by the sequential addition of characters, it builds a tree beginning with any two taxa and adding the taxa sequentially in such a way that the minimum number of evolutionary steps are employed (for a review with examples, see Wiley 1981a). This amounts to choosing the most parsimonious tree. The syntactical rule for making the most parsimonious decision involves the computation of path-length distances from each taxon added to the analysis and each ancestor already in the analysis. The most parsimonious solution is the one that results in the inclusion of all character information and that minimizes the amount of duplication of information already included in the tree to which the taxon is added. This amounts to rule 2 (i.e., $[R_{max}, S_{max}]$). The grouping rule for Wagner analysis differs from Hennigian phylogenetic analysis because it employs path-length distances. Farris and his colleagues have shown that the results of the algorithm are consistent with the results obtained by phylogenetic techniques employing rules 1 and 3. Thus, we may state:

$$\text{Rule } 4 = [(u_1, v_1), (u_2, v_2), \ldots, (u_n, v_n)]$$

where u_n = compute all possible path-length distances and v_n = add taxon to the branch that shows the smallest distance. Further, we can equate rule 3 with rule 4 because the path-length distances computed are based on synapomorphies and not on homoplasies or plesiomorphies. In fact, the Wagner algorithm can be implemented by hand without computing path-length distances if estimates of apomorphy are available (e.g., from outgroup comparisons). Thus, for Wagner analysis, the generative grammar takes the form

$$G: = [A_T, A_N, S_o, R_2, R_4]$$

Because we have shown that rule 1 and rule 2 are equivalent and that rule 3 and rule 4 are equivalent, both phylogenetics and the Wagner algorithm are compatible methods for phylogenetic inference under our theory.

Semantics

Once we leave syntactical structure considerations of our generative grammar and consider its semantic content, we bridge the gap between biological systematics per se and evolution. As suggested by our opening remarks, we do not mind doing this because maintaining such a disjunction is tantamount to not searching for a natural cause for the results obtained. Paraphrasing Papentin (1980), the elucidation of a pattern based on analysis of biological diversity without any information other than the pattern is only a hypothesis of the reality of the pattern; it says nothing about the laws underlying the reality of the pattern. Further, to fully explain why some techniques "work" and others are less satisfactory requires a theory of higher generality than the techniques themselves. A full justification for a systematic technique will remain in doubt unless coupled with a causal theory that explains its workings.

In formal language theory, such considerations come under the heading of justification of generative grammars. For example, phylogenetics can be viewed as a method for finding most parsimonious configurations of data. If it is found empirically that this is true, one may say that phylogenetics is justified. That is to say, it produces a consistent syntax. However, any systematic technique that performs its calculations accurately can be said to be equally well justified by the same criterion. This is the weakest of all justifications, because it does not allow us to choose one method over another as being most appropriate for a given purpose.

We would have a stronger justification if we advocated phylogenetic analysis because it provided the most information-dense (Farris 1979) or unambiguous (Brooks 1981a) summation of the data. This would allow us to express a preference for one method over another, but the method would still be justified by reference to criteria extrinsic to the entities being classified. Such justifications are *surface structure justifications* (fig. 5.1), or *external justifications,* or *measures of descriptive adequacy* (Chomsky 1965). Because linguists realize that no real language can be composed only of surface structure, generative grammars justified only in terms of surface structure are incompletely justified.

Generative grammars, or systematic methods in this case, may also be justified in terms of *deep structure,* or *measures of explanatory adequacy* (Chomsky 1965). Such justifications relate syntactical structure to the meaning of the message. Without explicit surface structure justification, there are no guidelines for semantic interpretation. Therefore, methods justified only in terms of deep structure are incompletely justified. Full justification of a systematic technique requires empirical justification of a consistent syntax based on some general principle such as information density or parsimony as well as consistency with a causal theory. And, as we have seen in the previous section, the dividing line between the two is often ambiguous. In the next section we will explore more fully the semantic component of the systematic method derived from our theory.

Semantic Components

There are two major semantic components involved in the systematics of our theory: (1) historical constraints on ordering of information and (2) minimum entropy of information. Since our theory stresses the importance of origins (the principle of individuality) and the historical burden of information carried by each species, we want our classifications to be genealogies based on maximum degree of historical constraint. We have shown that it is possible to generate a syntax from these components identical with the syntax of phylogenetic systematics. Both begin with a data base in which homologous characters are ordered into transformation series of primitive (plesiomorphic) and derived (apomorphic) characters. To do so requires an outgroup or an estimate of an outgroup, such as an ontogenetic transformation. Tree construction then proceeds by testing character distributions against all possible trees, and the preferred tree is the one that portrays the maximum amount of information with the fewest ad hoc hypotheses. What these procedures amount to is utilizing characters to arrive at a parsimonious solution of evolutionary descent. In this section we will attempt to show how the outgroup criterion and the implied outgroup criterion furnished by ontogeny relate to historical constraints and how parsimony relates to minimum entropy configurations.

Historical Constraints and Outgroups

To produce a hierarchy of species, we must be able to (1) order characters into plesiomorphic and apomorphic homologues and (2) maintain a positive divergence from equiprobability of possible trees. Outgroup comparison is sufficient to determine character polarity for homologous pairs of characters, but it does not necessarily result in a positive divergence from equiprobability of trees. In other words, there is no logical reason why the result would not be convergence toward equiprobability (i.e., random character evolution) unless the principle of historical constraints is working during evolutionary descent. We suggest, therefore, that the existence of hierarchies themselves are necessary and sufficient to demonstrate that historical constraints are operating. The outgroup criterion is successful in permitting us to reconstruct hierarchies only because historical constraints operate on the evolution of characters. The deep structure of the outgroup rule lies in understanding that ancestral information systems are not infinitely variable but inherently constrained relative to all conceivable future changes.

Parsimony and Minimum Entropy Increases

In practice, we commonly find ambiguities or character conflicts in data analyzed by phylogenetic methods. Parsimony is the method we use to resolve character conflicts when no other resolution is possible. But why? Do we use it because there is a deep structure link between the principle of parsimony and the evolutionary process? Or, do we use it because we have no other choice? If there is a biological reason to employ parsimony, then the principle is robust. If parsimony is employed simply because we cannot comprehend the complexities of the world, then we may never be able to deal with certain questions simply because they are masked by the need for simplicity. We suggest that there is a link between our theory and the reliance on parsimony in systematics, derived from the notion of minimum entropy increase.

Information Considerations

Parsimony is the criterion we use to decide which of two or more hypotheses is preferable. Parsimony concerns the economy or simplicity of one hypothesis relative to others (see Sober 1975; Beatty and Fink 1979 and references therein). In most contexts, parsimony has not been applied to processes but only to hypotheses concerning processes. The parsimony criterion is frequently used as a nonevidential criterion. For example, let us say that we are faced with the following situation: Sharks and bony fishes have lower and upper jaws, whereas lampreys and hagfishes do not. We might conjure up two possible explanations for the presence of jaws in sharks and bony fishes: (1) these characters evolved at different times from different ancestors and thus are not homologous or (2) these characters evolved in the common ancestor of sharks and bony fishes and thus are homologous. Vertebrate systematists have always opted for choice 2, principally because in the absence of any contrary evidence, it is more economical to conclude that such a complicated structure evolved once rather than twice. Using parsimony in this context results in maximizing redundancy. Thus, one aspect of parsimony can be directly linked to historical constraints minimizing entropy increases in information. We note that we do not have to assume that the evolution of characters is, in fact, parsimonious,

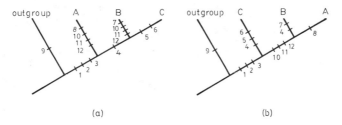

Fig. 5.4. Two phylogenetic trees summarizing the same characters. Tree *b* has a higher D_1 value than tree *a* and is more parsimonious.

since we do not have to assume that there are no homoplasies or that the production of homoplasies is impossible.

Another aspect of the parsimony criterion applied to systematics concerns the choice of particular phylogenetic trees when we have conflicting characters. This is an evidential criterion. In fig. 5.4, we have two hypotheses, both with support. The most parsimonious solution is fig. 5.4*b*, since to accept it we need postulate only a single instance of homoplasy (character 4), whereas if we accept fig. 5.4*a*, we must accept three instances of homoplasy (characters 10 to 12).

Using parsimony in this context amounts to maximizing D_1 (from rule 1). A second aspect of parsimony can be directly linked to minimizing entropy increases in information. The question now turns to whether or not evolution is actually parsimonious.

In a review of Sober (1975), Beatty and Fink (1979) suggested that evolution would be parsimonious if empirical findings resulted in two conclusions: (1) homoplasious characters are rare or (2) homoplasious characters occur at a lower frequency than homologous characters. The problem with a straightforward calculation of this lies with sampling error. We would need to know that the particular characters we use are a statistically random sample of all possible derived characters. We know of no way to ensure that this condition is met. However, there is another way to test the proposition. Consider a set of characters in which more than half the characters are homoplasious. There are two basic ways we might set up this data set. One way would be to have the homoplasies randomly distributed with respect to each other in such a way that the covariation among homoplasies is less than the covariation among homologues. In such a case, implementing a parsimony criterion would still result in a minimum entropy estimate of evolution, but evolution would have acted in a manner that resulted in a higher than minimally necessary entropy value for information.

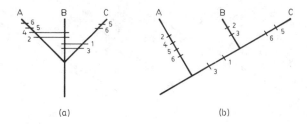

(a)   (b)

Fig. 5.5. Two phylogenetic trees summarizing the same characters. Search for the most parsimonious tree produces three equally parsimonious alternatives, which are summarized in a consensus tree (a). The actual phylogeny (b) is one of those three alternatives and not the consensus tree. The false trichotomy produced by parsimony analysis occurs because there are as many homoplasies as homologues and they all covary equally.

Consider another set of characters in which more than half the characters are homoplasious, but in this case consider that the co-variation among at least one set of homoplasies equals the covariation among the homologues. In such a case, one wrong tree and the correct tree will be produced when the parsimony criterion is implemented. The simplest example of such a result would be a trichotomous solution to a dichotomous tree, as shown in fig. 5.5. In this case, we have convergence on equiprobability, not divergence from equiprobability. If these characters represent the totality of character change during the evolution of ABC, then we may conclude that the minimum entropy principle is violated. Note, however, that the only way we might suspect this observation is through the use of the parsimony criterion.

We may also envision obtaining a dichotomous solution that is both unparsimonious and high on the relative entropy scale of character change. Consider a set of characters in which more than half the characters are homoplasious. Consider that the covariation of, for example, one set of homoplasies exceeds the covariation among homologies. The result of analyzing such a data set is shown in fig. 5.6 along with the true phylogeny. It is also possible to obtain a trichotomous solution when the true phylogeny is dichotomous. This could occur if an ancestor gave rise to two different peripheral isolates while not evolving itself. Such a situation, with the biogeography of the group, is shown in fig. 5.7; in these cases, historical constraints are not correlated with a series of speciation events. Note in this case that the minimum entropy configuration of information still obtains and that the solution is parsimonious.

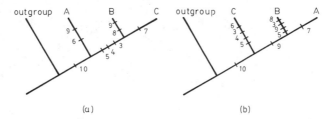

Fig. 5.6. Two dichotomous phylogenetic trees summarizing the same characters. Tree *a* is the most parsimonious, but tree *b* is the true phylogeny. The discrepancy arises because there are more homoplasious than homologous characters and the homoplasies all covary.

Finally, it is possible to obtain a false dichotomy for an evolutionary topology that should be trichotomous at the level of character analysis. Such a situation would occur if we had an ancestral species that did not evolve, budding off two peripheral isolates that later developed a shared homoplasy, as shown in fig. 5.8. In this case, an unparsimonious answer is the correct one.

Cohesion Considerations

The examples discussed above and graphically shown in figs. 5.5 to 5.8 are instances in which application of the parsimony criterion applied to information did not result in the correct phylogeny of the species involved. In figs. 5.5 and 5.6, this was due to homoplasy— a failure of historical constraints and thus a situation in which minimum entropy production of characters was not achieved during the

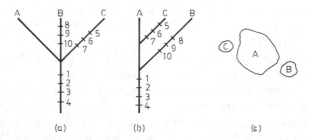

Fig. 5.7. Two phylogenetic trees summarizing the same characters. Tree *a* is the most parsimonious solution, whereas tree *b* is the actual phylogeny. The discrepancy arises because species A is a persistent ancestor, and B and C are descendants produced as peripheral isolates of A. *c,* The expected geographical distribution patterns for a persistent ancestor and two descendants produced as peripheral isolates.

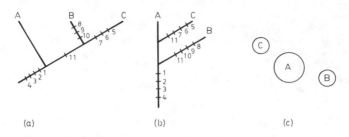

Fig. 5.8. Two phylogenetic trees summarizing the same characters. Tree *a* is the most parsimonious, but tree *b* is the actual phylogeny. The discrepancy arises because A is a persistent ancestor and B and C are descendants formed as peripheral isolates of A; furthermore, B and C each independently evolved character 11. *c*, The expected geographical distribution patterns for a persistent ancestor and two descendants produced as peripheral isolates.

evolutionary process. In fig. 5.5, the result was consistent with the true phylogeny but also consistent with an alternate false phylogeny. In figs. 5.6 and 5.8, the result was inconsistent with the true phylogeny. In fig. 5.7, the result was due to a lack of evolution in the ancestral species. We may see, therefore, that failure of the parsimony criterion for information occurs when the entropic change in information lags behind entropic changes in cohesion; that is, reproductive isolation occurs more often than novel characters evolve. This accords with our qualitative predictions in chapter 4 (fig. 4.17), in which the capacity that increased entropically at a slower rate was the determinate factor.

Testing for Departures from Minimum Entropy Configurations

We have considered how the deep structure of our theory relates to outgroup comparisons and the methodological parsimony criterion. We conclude that:

1. The occurrence of minimum entropy configurations involves entropy configurations of cohesion *and* information.

2. Minimum entropy configurations in phylogeny are related to the parsimony criterion but are not synonymous with it, because the parsimony criterion involves only information changes.

3. Methodological parsimony provides reasonable estimates of phylogeny, because there is a relationship between cohesion relationships and changes in information.

4. Parsimony breaks down only when character evolution is unconstrained to the point that the covariation of one or more sets of

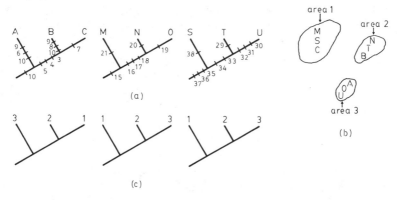

Fig. 5.9. Comparison of three phylogenies of different groups of organisms (a) inhabiting the same geographical areas (b). Two of the three phylogenies covary geographically and the third does not (c).

homoplasies exceeds the covariation of apomorphous characters. We do not expect this to occur due to historical and developmental constraints.

5. Parsimony will give an incomplete but consistent estimate of cohesion relationships when a clade exceeds the minimum entropy configuration of cohesion, as in the case of multiple peripheral isolates coupled with stasis in the ancestor or speciation via hybridization.

Given that evolution need not be characterized at each step by minimum entropy production in order to show a historical trend toward minimizing entropy increases, how would we test for departures from minimum entropy in particular cases? One strategy involves biogeographical analysis.

Species are found in biotas with other species. This fact gives us a chance to compare our estimates of the phylogenies of two or more clades by performing a vicariance analysis as discussed by Rosen (1978, 1979), Platnick and Nelson (1978), Nelson and Platnick (1981), and Wiley (1980b) (see also chapter 6). We can compare the phylogeny of our clade to other clades with similar distributions. In fig. 5.9a, the hypothesized relationships of the clade ABC and two other clades (MNO and STU) are shown inhabiting the same regions (fig. 5.9b). We might suspect that ABC is a problem because we encounter massive "homoplasy." Following the vicariance method, we check for congruence over time and space by substituting area names for species names in each phylogeny, producing area cladograms (fig. 5.9c). We find that the area cladograms for the clades MNO and

STU are congruent, but that the area cladogram for ABC is incongruent, inconsistent, and, indeed, opposite that of the other two.

There are many possible reasons why ABC's history looks different from the other two clades' histories. One of the reasons is that we have the wrong phylogeny; the reason we have the wrong phylogeny is that during evolution ABC exceeded the minimum entropy value that allows application of a parsimony criterion to give an accurate estimate of the genealogy. The potential for a test is therefore present. However, incongruence between area cladograms may be due to a number of factors other than true departures from minimum entropy production. Some of these factors are (1) dispersal and therefore different histories, (2) taxon extinctions, (3) different modes of speciation, and (4) examining the wrong set of characters, thereby biasing the results toward high-entropy configurations when adding other characters would lead to a minimum-entropy phylogeny.

We conclude that while the potential for a test exists, many alternatives exist for explaining incongruent patterns and the investigation must proceed with caution.

"Linguistic Affinities" of Systematic Techniques

The systematic technique favored by many contemporary evolutionary biologists is, like neo-Darwinism itself, a diverse pluralism sharing a common name. In this case, it is called *evolutionary taxonomy* (Mayr 1969). There has not been as explicit a methodological statement made about evolutionary taxonomic protocols as there has been about phylogenetics. However, the theoretical underpinnings of evolutionary taxonomy center around the assertion that phylogeny is a mixture of genealogical and adaptive relationships (e.g., Mayr 1981). Without having to prescribe precise formal operations, we may nonetheless see that two sets of rules are necessary to implement this approach. One set of rules applies to delineating genealogy, and the other applies to placing parts of the genealogy in the proper adaptive or anagenetic position. We may symbolize this as R_5, or rule 5.

$$R_5 = [(u_1,v_1),(u_2,v_2),...,(u_n,v_n)]$$

where u_n = delineate genealogy and v_n = group species in proper adaptive or anagenetic context. Since u_n is a strictly genealogical criterion, we may equate it with the rules for phylogenetic analysis discussed earlier. Thus, evolutionary taxonomy, as a generative grammar, can be summarized thusly:

$$G: = [A_T, A_N, S_o, R_s]$$

Because, as we have shown, Hennigian argumentation and the Wagner algorithm share a common syntax, they may be considered idiolects of the same language. Churchland (1979) discussed a concept called *homodoxy,* or sameness of individual understanding across idiolects. Two terms, T_x and T_y, are homodoxical if, and only if, they are paired in an optimal transdoxation of one idiolect to another. And, for idiolects sharing a common syntax, an optimal transdoxation of idiolect I_x into idiolect I_y is a mapping of the terms of I_x into the terms of I_y such that, for each term T_x, images under that mapping of the systemically more important of the I_x sentences are exactly the same as the systemically more important of the I_y sentences containing its paired term T_y. For biological classifications, the terms referred to would be the terminal taxa and the supraspecific groupings indicated by the classification. Given a set of data, both methods of phylogenetic analysis will produce classifications in which all paired terms are homodoxical and for which any unique terms will be consistent (*sensu* Wiley 1981b) with the homodoxical terms.

The same cannot be said to be true for classifications produced by evolutionary taxonomists; that is, the results of evolutionary taxonomic analyses are not always consistent with those of phylogenetic systematics. Thus, linguistically, evolutionary taxonomy and phylogenetic systematics are not idiolects, but different languages, even though they use the same empirical data base and appear to use the same syntactical structure. It is possible, however, that the two languages represent the same causal theory, that is, have the same deep structure. A formal way to ascertain this is to translate one into the other. Churchland's (1979) view of translation was this: The aim of translation is to find, to the extent that it is there to be found, an intensional (deep) structure in the target language that parallels the intensional structure of one's own. Faithful translation between languages is possible only to the extent that the two languages involved do, indeed, have closely similar intensional structures. Or (Churchland, 1979):

An optimal translation of L_a into L_b is a mapping of the terms *and syntactic forms* of L_a into the terms *and syntactic forms* of L_b such that (1) the image, under that mapping, of each grammatical sentence of L_a is a grammatical sentence

Table 5.7   Evolutionary Taxonomic Classification of Major Groups of Amniotic Vertebrates and Phylogenetic Systematic Classification of the Same Groups

Evolutionary taxonomic classifiction	Phylogenetic classification
Ammiota	Amniota
Reptilia	Chelonia
Chelonia	Mammalia
Crocodilia	Archosauria
Rhynchocephalia	Aves
Squamata	Crocodilia
Sauria	Diapsida
Ophidia	Squamata
Mammalia	Sauria
Aves	Ophidia

of L_b, and (2) the image, under that mapping, of the sentences semantically important for each term T_a in L_a are exactly the sentences semantically important for its paired term T_b in L_b.

Under this view, two terms, T_a and T_b, are synonymous if, and only if, they are paired in an optimal translation of one language into the other.

As an example, consider the translation of an evolutionary taxonomic classification of the amniote vertebrates into a phylogenetic classification of the same group (table 5.7). It is possible to make grammatical "sentences" using all of the basic terms, but some of the higher level terms have been lost as a result of the syntactical changes required. Other terms would be lost in a translation from the phylogenetic classification into the evolutionary taxonomic one. Most important, however, is the recognition that the "sentences" lost in each case are semantically important. Reptilia are semantically important to evolutionary taxonomy as evidence for an adaptive zone and adaptive radiation. Archosauria (birds + crocodilians) are semantically important to phylogeneticists as evidence for common ancestry. This suggests that the intensional (i.e., deep) structure of the two systematic languages is different.

We know the intensional structure (the properties connoted by a term) of phylogenetics: (1) *species* are individuals, (2) *groups of species* are strictly monophyletic, and (3) *evolutionary history* is always genealogical and may be adaptive as well. This can be deduced from our causal theory. What about evolutionary taxonomy?

We do not think the question has ever been seriously considered, because evolutionary taxonomy has never had a structuralist framework. It has never been called upon to play an active role in evolutionary theory. However, its intensional structure is avowedly neo-Darwinism (see Mayr 1969). Based on our analysis of neo-Darwinian systematic theory (chapter 1), we may assert an intensional structure consistent with those aims: (1) *species* are natural kinds, or classes; (2) *groups of species* may be monophyletic or paraphyletic, depending on (3) *evolutionary history,* which is always adaptive and may be indicative of genealogy in a general manner. Clearly, the intensional structure of the two approaches differs. And if this is true, can they really be derived (or derivable) from the same causal theory? We have already said that it is the causal theory from which a generative grammar derives its deep structure justification. Phylogenetics and evolutionary taxonomy certainly connote different theories about the origin of higher taxa and the explanation for long-term evolutionary trends. And, without an explanation for the natural hierarchy and those long-term trends, neo-Darwinism is not a complete theory, regardless of its population genetic basis. By highlighting the intensional differences between the two approaches, we can readily see that their apparent similarities in syntactical structure are not real but merely functional equivalents. Evolutionary taxonomic classifications are passive repositories of descriptive information gathered under one theory, whereas phylogenetic systematic classifications are analytic statements derived from a different theory.

Summary

It has been possible to derive a formal technique for reconstructing phylogenies, and deriving classifications from them, from the basic principles of our theory. That systematic technique is compatible with various forms of phylogenetic systematics, a technique already in use by a growing number of systematists. Historical constraints in evolution even at this functional level produce a trend toward minimizing entropy increases. A formal language theory analysis suggests that our theory is not compatible with evolutionary taxonomy and, by extension, with some of the evolutionary explanations drawn from such an approach. The evolutionary explanations that conflict are those involving adaptive groups that are manifestations of supra-specific evolution, with which we took exception in chapter 1.

In chapter 6 we will investigate some ways in which the results of systematic analysis can play a crucial role in understanding the evolution of ecological associations.

· 6 ·

Historical Ecology

The highest functional level of biological macroscopic order commonly studied is that of ecological associations such as communities, ecosystems, or predator-prey, host-parasite, and herbivore-crop interactions. This involves not only the relationship between organisms and their abiotic environment, but also the apparent functional "fit" of organisms to each other. Under functionalist, or top-down, theories of biological organization, this would be the level from which explanatory analyses would begin. Important capacities identified at the level of communities would be considered the primary causal agents of organization at the next lower level, that of species and populations. Our approach is structuralist, or bottom-up, and explains higher levels of organization as emergent properties of lower levels. This is not a reductionist view, since we believe that communities exhibit properties not found in lower functional levels of biology. Despite our heterodox approach, we will try to explain properties of interest elucidated by previous studies. In this chapter, we intend to show that (1) the origin and evolution of ecological associations can be understood as a nonequilibrium entropic phenomenon, (2) much work that has already been done by other workers can be integrated under a unified theory, and (3) previous attempts at achieving a real synthesis have been hampered by the lack of a protocol for discerning historical components in the evolution of ecological associations.

If ecological associations are nonequilibrium entropic phenomena, we would expect them to be energy-processing systems and to exhibit self-organizing capabilities. In chapter 4, we suggested that sexually reproducing species could be considered energy-processing systems because some of the energy taken up by some members of the species, the parents, could be retained within the species, as offspring, longer than the life span of the parents. In an analogous manner, any ecological association is guaranteed to have some energy-processing capability if energy taken up by one member of the association is transferred, at least in part, to one or more other members. We have also noted in previous chapters that energy flows are highly asymmetrical in nonequilibrium systems; that is, they are predominantly one-way. In ecological terms, this means that the evolution of heterotrophic organisms inevitably led to the existence of ecological associations as nonequilibrium systems, at least in some minimal sense. Furthermore, the greater the degree of asymmetry in energy flows among members of an association, the more pronounced should be the nonequilibrium nature of the association. Nonequilibrium systems, as we have shown, exhibit organization proportional to the difference between the maximal possible entropy and the actual entropy observed given the components at any time.

Our theory also suggests that ecological associations are self-organizing systems. One way in which the self-organizing capabilities of an ecological association can be assessed is by estimating the degree to which the membership and energy flow pathways of an association are historically determined. The greater the proportion of historically determined components, the greater the self-organizing capabilities.

In providing an evolutionary explanation for any ecological association, it appears that two general components must be included; first, an estimate of historically determined components and, second, an estimate of the relative entropy and organization of the system. This chapter outlines an experimental protocol for obtaining some assessment of those parameters.

Individuals belonging to different species inhabiting the same area form *associations;* if these individuals interact with each other they form a *community* (Ricklefs 1979). The extent to which associations and communities are the product of congruent patterns of descent indicates the extent to which the association or community is a product of a common history and, in the case of a community, a product of a history of ecological interaction. The extent to which associations and communities have component species that do not share a congruent pattern of descent is an indication of the extent to which unique historical factors, primarily dispersal and extinction, have entered into considerations of community structure or faunal or floral associations. In this chapter, we are interested in two basic areas. First, how can we sort out the relative contributions of dispersal and history to our understanding of the present structure of biotas and communities? Second, how can we apply entropy measures to particular communities? The second objective is meant to integrate our considerations of the entropic nature of speciation and populational phenomena with our view that communities themselves can be entropic systems.

Life History Cycles and Ecology

The life history cycle of an individual includes its ontogeny and its interactions with its environment. This life history cycle includes historical components derived from the individual's parents and the geographical place in which the individual finds itself, and proximal components, the particular and frequently changing abiotic and biotic components of its environment. The historical components constrain the repertoire of responses by an individual to its environment, whereas the environmental (proximal) components determine which of these constrained responses will be manifested at a particular time.

We might think of a species as having certain ecological attributes manifested by its members. Since communities are trophically structured, and particular trophic levels are universal class constructs, we expect that different species may show many similarities in their ecology. The class "acid bog–dwelling autotroph" is likely to be filled anywhere there is an acid bog.

A species may have a particular ecology because (1) that ecology is the same as the ecology of its ancestor or (2) that ecology is a modified version of the ancestral ecology. In the first case, the question of why the species has a particular ecology is a historical question involving not the particular species per se, but its immediate ancestor and perhaps other ancestors in the past. For example, at least nine species of the topminnow genus *Fundulus* are typical "insectivores" that pick insects from the surface of the water. Such feeding behavior is widespread in the order Cyprinodontiformes to which these fishes belong. Obviously, explaining the origin of this ecology as being peculiar to, say, *Fundulus notatus* because it allows *F. notatus* to survive in communities where it is found would not be an adequate explanation. *F. notatus* is doing a very ancient thing. At least some of its ecological life history traits are not the product of the communities in which its members are now found but of very ancient origin about which we may be largely ignorant. The particular ecology of *F. notatus* is analogous to a primitive morphological character. However, some of the structure of communities in which *F. notatus* occurs cannot be ascribed to this analogue of phylogenetic primitiveness. *F. notatus* is broadly sympatric with a close relative, *F. olivaceous*. They eat the same food (Thomerson and Wooldridge 1970) and largely prefer the same habitat. But, they are largely "allotopic" (not found at the same site together), a distributional pattern Thomerson (1966) ascribed to competitive exclusion. Unlike the primitive aspects of the ecology of *F. notatus* and of *F. olivaceous*, the observation of allotopy requires a proximal cause that can be inferred and studied without recourse to historical data. Thus, certain aspects of the ecology of this species require historical explanation, whereas others are amenable to proximal explanations.

Coevolution and Colonization

Ecological associations and communities represent higher levels of biological organization above that of species. Like species, ecological associations and communities have properties (analogous to characters) that may be the result of history or of proximal causes. Unlike species, neither ecological associations nor communities have intrinsic boundary conditions analogous to, for example, reproductive isolation. A biota may be defined as an array of ecological associations composed of one to many communities whose boundaries are determined by physiographic and climatic parameters. Some biotas are easily defined because they form a single ecological association involving species not found in other areas. Such biotas are termed *areas of endemism* by some biogeographers (e.g., Platnick and Nelson 1978). Other biotas are composed of many ecological associations which, as a whole, share common ecological properties but are not composed of species that always co-occur. For example, the eastern deciduous forest of North America shows a change in tree associations throughout the biota; no one species "defines" the area nor is the area an area of endemism (although it contains sub-areas of endemism). Nevertheless, it seems to be a biota because we know that it is a historical subunit of a formerly widespread hardwood forest whose remnants are scattered from Guatemala to China. A biota may have a strong historical component implying its "descent" from an ancestral biota as well as a proximal component relating to current physiographic and climatic conditions. Indeed, the extent to which we can recognize a biota at all may be a function of the historical component it contains and a recognition that some species within the biota are closely related to species occurring in similar and presumably related biotas.

Separating historical and proximal causes for ecological associations involves at least two components. First, we may ask what climatic and physiographic parameters correlate with the particular association we observe. Such analyses provide a partial answer to why associations may change over space or through time. Second, we may ask which species in a particular ecological association occur together because their ancestors occurred together and which represent colonization events, that is, which evolved in different areas or perhaps different biotas and dispersed into the area now having the particular association we observe. Put simply, is the particular ecological association a historical association, an association brought about by colonization, or a combination of the two? We expect it to be a combination of the two for dispersal is a fact of life. But what kind of combination? Are most ecological associations maintained as equilibrium systems fueled by constant dispersal or are they historical self-organizing associations?

One way to approach the question is to study the co-occurrence of monophyletic groups. There appears to be a strong historical component to the association in some cases studied. For example, McCoy and Heck (1976) pointed out that the ecological association between hermatypic corals, mangroves, and seagrasses has a strong historical component, probably Cretaceous in origin. This is reflected in the distributional histories of these groups. There is also a strong ecological component, equally ancient, reflected in the succession of these communities (Welch 1963). The corals colonize a hard substrate, permitting filamentous algae to move in. Mat-forming calcareous algae succeed the filamentous algae and provide a substrate for the seagrasses. The seagrasses build up the substrate, permitting the invasion of mangroves that shade out the seagrasses. One might assume from this description of ecological succession that the circumtropical distribution of mangrove communities is a dispersal phenomenon. It certainly was at some level, but this dispersal apparently occurred before the dispersal of the major continental land masses (McCoy and Heck 1976, 1983).

Food webs may also have historical components. Domning (1981) has pointed out that the association of seagrasses and sea cows is extremely ancient. Sea cows have, apparently, grazed on seagrasses since the origin of sea cows in the early Eocene. Further, the presence of seagrasses is correlated with a distinctive invertebrate and foraminiferan fauna that permits predictions concerning the occurrence of seagrasses in paleocommunities where they would not be preserved (literature summarized in Domning 1981). Thus the recognition of a historical component in present ecological communities permits a rich array of predictions concerning the distribution of these communities in time and space.

Vicariance Biogeography

Another way that we might separate historical and proximal causes that explain ecological associations and community structure is to apply the methods of vicariance biogeography developed by Croizat, Nelson, and Rosen (1974), Platnick and Nelson (1978), and Rosen (1978) and summarized by Nelson and Platnick (1981) and Wiley (1980b, 1981a). The basic strategy is to reconstruct the phylogenies of two or more groups of organisms and to compare the histories of these clades with respect to phylogenetic and biogeographical congruence. Essentially, we ask whether these clades show a pattern of consistency. If so, their ecological association is probably historical. If not, their ecological associations are the result of colonization of an ecologically suitable habitat. When we say that the phylogenies are consistent, we do not imply that they are exactly the same. Consistency is different from congruence. Two trees that are fully congruent with respect to speciation and biogeography may denote a common pattern of *cospeciation* (Brooks 1979). Cospeciation is an aspect of coevolution in which speciation of the ancestors of two or more clades are coincident in time and space. A lack of speciation of one of the clades produces a pair of phylogenies that is not fully congruent but is consistent, which does imply a historical association. Inconsistencies imply colonization. Sorting out congruent, partly congruent (consistent), and incongruent (inconsistent) patterns of evolution in time and space is a problem that can be at least partly solved with vicariance methods.

Vicariance biogeographers have adopted the following tenets to guide their analyses.

1. The first phase of analysis should be a data-collecting phase. The distribution of various natural groups should be plotted on maps. Such plots are termed *tracks*. A search should be made for replicated patterns of distribution (termed a *generalized track*). This phase may be termed *track synthesis,* and it is a basic data-reduction phase (Croizat 1962).

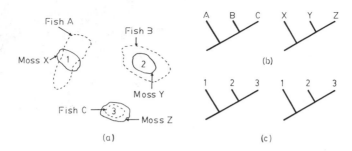

Fig. 6.1. Biogeography of hypothetical fish and moss taxa. a = Ranges of the subtaxa in each group; b = phylogenetic relationships of the subtaxa; c = area cladograms of the two groups showing a common history. (Redrawn from Wiley, *Syst. Bot.* 5 [1980]:194–220.)

2. From the data collected in the track synthesis stage, areas of endemism are identified. *Natural sets of endemic taxa* (i.e., putative historical ecological associations) may then be analyzed with respect to the phylogenetic interrelationships of their respective floras and faunas. Two questions are asked: What are the interrelationships among the organisms inhabiting the different areas of endemism? And, how do those relationships relate to the geographical and geological histories of the areas themselves?

3. To help answer these questions we need phylogenetic hypotheses of the organisms inhabiting the areas in question.

4. From an array of phylogenetic hypotheses we search for congruence in relation to areas inhabited. To accomplish this, the area in which a species lives is substituted for that species name in the phylogenetic hypothesis. Such a hypothesis has been termed an *area cladogram*. For example, in fig. 6.1, we see two groups, fish and moss, confined to three areas of endemism, 1, 2, and 3. The phylogenetic relationships of fish and moss are analyzed. The area of each species is then substituted for the species name (i.e., substituting "1" for "A" and "X," etc.; fig. 6.1c). In this example, there is complete congruence between the hypothesized phylogenetic relationships of each group and the areas in which they occur. We may conclude that cospeciation is a likely explanation for these two groups and that the ecological association is a historical one.

5. The frequency with which we find complete congruence or consistency is related to the frequency with which *common* or *general* factors affected the evolution and distribution of two or more groups of organisms. To search for such congruence we must separate unique factors particular to one group from common factors

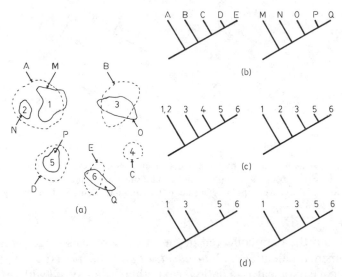

Fig. 6.2.   Biogeography of two hypothetical groups (A-E) and (M-Q). *a* = Ranges of the subtaxa of each group; *b* = phylogenetic relationships of the subtaxa; *c* = area cladograms of the groups (note the area cladograms are not fully congruent); *d* = reduced area cladograms of the two groups showing components of common history. (Redrawn from Wiley, *Syst. Bot.* 5 [1980]:194–220.)

related to the evolution of all groups considered. Rosen (1978, 1979) accomplished this by deleting unique factors from the area cladogram. For example, fig. 6.2 shows the distributions of two monophyletic groups (A-E and M-Q) among six areas (1 to 6). The phylogenetic relationships of these groups are shown in fig. 6.2*b*. The area cladograms are shown in fig. 6.2*c*. Note that the area cladograms are not fully congruent but that the phylogenies of both groups are consistent. For example, areas 1 and 2 are both inhabited by species A and by two different species of group M-Q (M in 1 and N in 2). Also, area 4 is inhabited by C but not inhabited by any member of M-Q. We attempt to find the congruence between these two groups and the areas they inhabit by deleting the unique factors of each group. This produces what Rosen (1978) termed a *reduced area cladogram* (fig. 6.2*d*). These reduced area cladograms show congruence because the areas of one group are arranged exactly like those of the other group. This is the level at which the histories of the two groups show a history of cospeciation. However, this does not mean that the incongruent elements show a lack of historical association. Species A is historically associated with both species M and N because species A was associated with the ancestor of M and N and no colonization need be postulated. The lack of association of species C with a member of the M-Q clade, however, does require an explanation, either that C colonized area 4 or that a species of clade M-Q (or a population of a species) became extinct in area 4.

6. Reduced area cladograms imply that congruence observed is due to factors that affected both groups. The next step would be to search for a geographical cladogram, which might detail the relationships of the various endemic areas without reference to the organisms that inhabit those areas. Such geographical cladograms (or maps that can be easily converted) are common on the level of continental drift, but few have been produced for smaller geographical areas.

Applications of this method have been few so far, but the results are sufficiently interesting to pursue further studies. Rosen's (1979) analysis of two genera of live-bearing topminnows (*Heterandria* and *Xiphophorus*, family Poeciliidae) indicate that the Middle American species of *Xiphophorus* commonly termed swordtails share at least four congruent speciation events with species of *Heterandria*. The incongruent events in both clades may be due to the fact that peripheral isolation caused speciation independently in each genus (an interpretation made by Wiley 1980b). Wiley (1981a) concluded that vicariance and peripheral isolation might result in different phylogenetic patterns; specifically, that while cospeciation might be expected when biotas are subdivided, peripheral isolation may lead to speciation only in particular clades. Thus the method has possibilities for explaining the presence of faunal or floral associations and also possibilities for explaining consistent but incongruent patterns.

Consider fig. 6.1 again. The phylogenies of the associates "fish" and "moss" are highly correlated with each other relative to the areas in which particular members occur. We may combine the implied geographical information from each phylogenetic tree in a single summary branching diagram representing geographical relationships. To do this, we convert each phylogenetic tree into a matrix of binary codes (fig. 6.3*a,b*) and use them as "characters" of the three areas (fig. 6.3*c*). The resulting tree (fig. 6.3*d*) shows perfect fit of both groups of organisms to a single tree. From this we can "read" that A and X, B and Y, and C and Z are associated because of common history. We expect the phylogenies of members of associations which have evolved together to be highly correlated with each other and with the history of the areas in which they

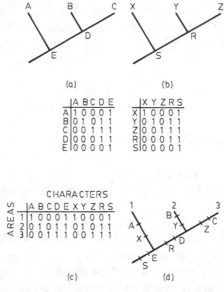

Fig. 6.3. Quantitative method for assessing degree to which two different mono-phyletic groups share a common geographical history. *a,* Cladogram of taxon "fish" from fig. 6.1b with interior nodes labeled (*top*) and converted into a matrix of binary characters (*bottom*). *b,* Cladogram of taxon "moss" from fig. 6.1b with interior nodes labeled (*top*) and converted into a matrix of binary characters (*bottom*). *c,* Matrix of binary characters symbolizing taxa fish and moss arranged according to geographi-cal distribution in areas 1, 2, and 3. *d,* Cladogram of relationships among areas 1, 2, and 3 based on phylogenetic relationships of taxa fish and moss. Note con-gruence.

occur. Phylogenies of species which have colonized a given as-sociation will not be so highly correlated (see also Brooks 1981b).

In fig. 6.4, we have added a new clade (JKL), which also has members in areas 1, 2, and 3, but in a different pattern (fig. 6.4*a*), than the historical one. By adding the binary codes for JKL to the data matrix (fig. 6.4*b*), we obtain the summary tree shown in fig. 6.4*c*. Note that the historical structure of the associ-ations is the predominant feature. However, taxon J occurs more than once on the cladogram. Since a species cannot evolve more than once, we know that J has colonized at least one of the areas. In this case, since K also occurs in area 2, in accordance with the overall historical relationships, we can postulate that J colonized area 2.

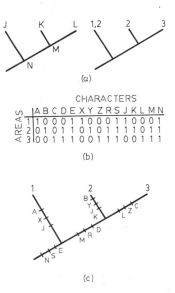

(a)

CHARACTERS

	A	B	C	D	E	X	Y	Z	R	S	J	K	L	M	N
1	1	0	0	0	1	1	0	0	0	1	1	0	0	0	1
2	0	1	0	1	1	0	1	0	1	1	1	1	0	1	1
3	0	0	1	1	1	0	0	1	1	1	0	0	1	1	1

AREAS (1, 2, 3 on the left side)

(b)

(c)

Fig. 6.4.   Example from fig. 6.3 with third taxon (JKL) added to analysis. *a*, Clado-gram for taxon JKL with interior nodes labeled (*left*) and with geographical distri-butions replacing taxon designations (*right*). *b*, Matrix of binary characters symbolizing taxa fish, moss, and JKL arranged according to geographical distribution in areas 1, 2, and 3. *c*, Cladogram of areas 1, 2, and 3 based on phylogenetic analysis of taxa fish, moss, and JKL. Note that taxon J occurs in area 2 along with taxon K, in addition to occurring in area 1, its historically predicted site. This indicates dispersal by J from area 1 to area 2.

Separating Historical and Proximal Ecological Associations—Two Examples

For one example, we have chosen the helminth parasites of neo-tropical freshwater stingrays (family Potamotrygonidae). All of the helminths studied have rather specific, and different, invertebrate intermediate hosts as well as their stingray definitive hosts. Thus, each parasite is really a symbol for at least two invertebrate and one vertebrate species and represents a portion of a larger community including nonparasitic organisms. Phylogenies of these parasite taxa have been published elsewhere (Brooks, Mayes, and Thorson 1981; Deardorff, Brooks, and Thorson 1981), as have their geographic distributions (Brooks, Thorson, and Mayes 1981). Fig. 6.5 is the cladogram of areas based on those parasites. Using this cladogram, we may interpret the history of each part of the community (parasite species or lineage in this case) across all communities in which they occur. We may then produce statements about the evolution of each

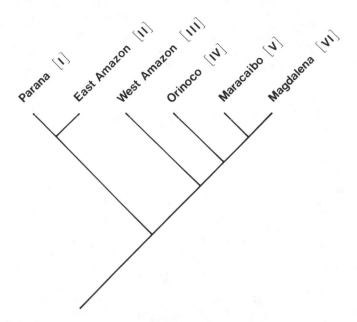

Fig. 6.5.   Cladogram for six neotropical areas based on helminth parasites inhabiting freshwater stingrays. (Redrawn from Brooks 1985.)

member of each community. For example, for a given geographical area, do the parasite communities with definitive stingray hosts show a pattern of colonization or coevolution with their hosts? We can approach this question by seeing if we must postulate dispersal to explain how these species have attained their present distribution.

   Acanthobothrium (fig. 6.6). This group of tapeworms shows the closest correlation with the overall cladogram topology. *Acanthobothrium quinonesi,* occurring in both the Magdalena area and the Maracaibo area, acts like a synapomorphy linking those two areas. This may be a result of differentiation of *A. quinonesi* in an area comprising both the Magdalena and Maracaibo, with subsequent subdivision into two allopatric populations. Or, it may be the result of differentiation of *A. quinonesi* in one area (probably the Magdalena area) and dispersal into the other (probably the Maracaibo area) (see Brooks, Thorson, and Mayes 1981).

   The absence of *Acanthobothrium* in the East Amazon area may be due to sampling error or of the failure of the lineage to disperse into the area from the south. If the result of a sampling error, the species to be found is predicted to be either *A. terezae* or its sister species. This can be tested by reference to the character-state dis-

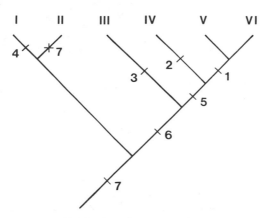

Fig. 6.6. Character-state distribution of members of *Acanthobothrium* (Cestoda) in areas I to VI. Slash marks numbered 1 to 4 represent *A. quinonesi* (*1*), *A. regoi* (*2*), *A. amazonensis* (*3*), and *A. terezae* (*4*). Slash marks numbered 5 to 7 represent hypothetical common ancestors of 1 + 2 (*5*), 3 + 5 (*6*), and 4 + 6 (*7*). X-numbered 7 refers to postulated loss of *Acanthobothrium*, which could also indicate sampling error. (Redrawn from Brooks 1985.)

tributions presented in the cladogram of this group of species (Brooks, Mayes, and Thorson 1981).

Rhinebothroides (fig. 6.7). This genus of tapeworms represents the most highly diversified group of parasites known from potamotrygonids. The phylogeny of *Rhinebothroides* represents two trends (fig. 6.7*a,b*). The first parallels the pattern shown by *Acanthobothrium* (fig. 6.6). It represents a vicariant pattern involving the Parana–West Amazon–Orinoco–Magdalena areas. *R. venezuelensis* occurs in the Maracaibo area as well as the Orinoco and results either from a dispersal event or from a previous connection of the two severed areas, leaving two allopatric populations.

The second pattern (fig. 6.7*b*) involves an alternative pattern of areas—Parana–East Amazon–Orinoco. The node connecting this group of species to the rest of the *Rhinebothroides* cladogram also serves as the Parana component for the pattern shown by the other species. Because of that, and because *R. scorzai* occurs in the Parana and Orinoco (and undoubtedly in the East Amazon, although it has not yet been collected), there appears to have been the kind of local differentiation and secondary dispersal one associates with a "peripheral isolate" model of speciation (Wiley 1981a).

Potamotrygonocestus (fig. 6.8). The three known species of this cestode genus exhibit a distribution pattern (fig. 6.8) like that of *Acanthobothrium* (fig. 6.6) or part of *Rhinebothroides* (fig. 6.7),

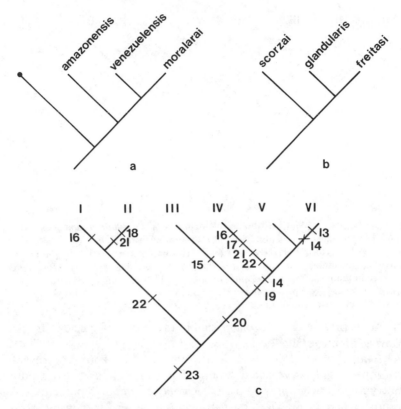

Fig. 6.7. Character-state distribution of members of *Rhinebothroides* (Cestoda) in various areas. Fig. 6.7*a* represents one part of cladogram, symbolized in fig. 6.7*c* by *13* (*R. moralarai*), *14* (*R. venezuelensis*), *15* (*R. amazonensis*), *22* (an ancestral node), *19* (ancestor of 13 + 14), and *20* (ancestor of 15 + 19). Fig. 6.7*b* represents the second part of the cladogram, symbolized in fig. 6.7*c* by *16* (*R. scorzai*), *17* (*R. glandularis*), *18* (*R. freitasi*), *21* (ancestor of 17 + 18), and *22* (ancestor of 16 + 21). Fig. 6.7*a,b* join at 22.

except that there is no evidence for even an inferred ancestor in the Parana region. We suspect there is an endemic species of *Potamotrygonocestus* in the Parana. Like *R. scorzai*, *P. amazonensis* is a widespread species, occuring vicariantly in the West Amazon area and, apparently by dispersal, in the East Amazon, Orinoco, and Maracaibo.

The most efficient map of the *Potamotrygonocestus* species on the summary cladogram (fig. 6.5) is shown in fig. 6.8*b*. Because such a set of relationships does not correspond well to a genealogy based on observed attributes of the species involved, we interpret fig. 6.8*b*

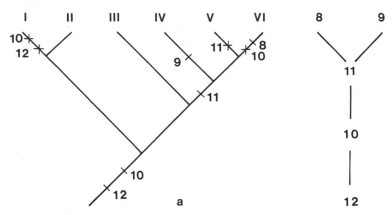

Fig. 6.8. Character-state distribution of members of *Potamotrygonocestus* (Cestoda) in various areas. *a*, Distributions of *P. magdalenensis* (*8*), *P. orinocoensis* (*9*), *P. amazonensis* (*10*), the ancestor of 8 + 9 (*11*), and the ancestor of 10 + 11 (*12*). X denotes postulated secondary loss if vicariance is the only explanation. *b*, The most parsimonious ordering of *Potamotrygonocestus* based on observed distributions. Lack of congruence between cladogram and 6.8*b* is interpreted as evidence of widespread dispersal and/or sampling error. (Redrawn from Brooks 1985.)

as an indication of the widespread dispersal exhibited by *P. amazonensis*.

Ten taxa found in these areas show no differentiation (fig. 6.9). These ten taxa fall into two groups: (1) those acquired secondarily by host-switching from local teleosts and (2) those occurring in undifferentiated form that were apparently original constituents of the helminth fauna of the ancestral potamotrygonid first isolated in South America (see Brooks, Thorson, and Mayes 1981).

Echinocephalus daileyi occurs in the West Amazon, Orinoco, and Parana (the latter documented by A. A. Rego, Instituto Oswaldo Cruz, pers. comm.). It is therefore consistent with the vicariance pattern shown by *Acanthobothrium*, part of *Rhinebothroides*, and *Potamotrygonocestus* and represents part of a historical association that has many species.

Eutetrarhynchus araya and *Rhinebothrium paratrygoni* occur in the Parana and Orinoco; *E. araya* also occurs in the East Amazon. These two species seem to exhibit the same pattern as shown by part of *Rhinebothroides* (fig. 6.7*b*) and thus represent part of the historical association. Apparently the isolating events that resulted in differentiation of three species of *Rhinebothroides* did not have the same effect on *R. paratrygoni* or *E. araya*.

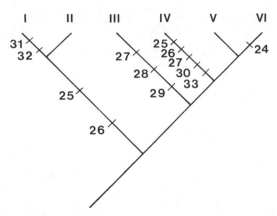

Fig. 6.9. Character-state distribution of undifferentiated helminth parasite taxa in freshwater stingrays. *24* = *Paravitellotrema overstreeti* (Digenea); *25* = *Eutetrarhynchus araya* (Cestoda); *26* = *Rhinebothrium paratrygoni* (Cestoda); *27* = *Echinocephalus daileyi* (Nematoda); *28* = *Potamotrygonocotyle amazonensis* (Monogenea); *29* = *Heteronchocotyle tsalickisi* (Monogenea); *30* = *Terranova edcaballeroi* (Nematoda); *31* = *Leiperia gracile* (Pentastomida); *32* = *Brevimulticaecum* sp. (Nematoda); *33* = *Megapriapus ungriai* (Acanthocephala).

The two species of monogeneans (*Potamotrygonocotyle tsalickisi* and *Paraheteronchocotyle amazonensis*) found in the West Amazon represent curious relics. Monogeneans have not been found on any other potamotrygonids.

Finally, *Paravitellotrema overstreeti, Megapriapus ungriai, Leiperia gracilis, Terranova* sp., and *Brevimulticaecum* sp. all are localized species normally occurring in non-stingray hosts. These species were not part of the historical association and represent added ecological interactions within the community of potamotrygonid parasites.

We can summarize our results in a series of statements about the stingray parasite communities in each of the six areas studied.

Parana. There is no indication of colonization of Parana rays by parasites differentiated in other areas. Nor do we know of any local parasites that have expanded their habitats to include potamotrygonids. Of the six communities identified, this appears to be the earliest one differentiated (see Brooks, Thorson, and Mayes 1981). Subsequently, there appears to have been a good deal of dispersal from the Parana to the East Amazon to the Orinoco following differentiation of the Parana fauna, but little dispersal into it.

West Amazon. This area also shows no indication of colonization of rays from other areas or from local parasites normally associated with non-stingray hosts. It contains two species, *Echinocephalus daileyi* and *Potamotrygonocestus amazonensis,* which are widespread in distribution. The Parana community is its closest, and apparently vicariant, relative.

Magdalena. With the exception of *Acanthobothrium quinonesi,* which also occurs in the Maracaibo area, all known parasites found in stingrays in this area are endemic. This includes *Paravitellotrema overstreeti,* which represents a group of digeneans normally parasitizing siluriform and characid teleosts. The other endemic parasites apparently vicariated in the area, with *A. quinonesi* dispersing into the Maracaibo area.

The three preceding communities appear to be primarily vicariant communities, some of whose members have dispersed into other areas. The next three communities present more interesting histories.

Maracaibo. This area has only a single, endemic, species of potamotrygonid (*P. yepezi*). That stingray is known to host three species of helminth parasites, *Acanthobothrium quinonesi* from the Magdalena fauna, *Potamotrygonocestus amazonensis* from the West Amazon fauna, and *Rhinebothroides venezuelensis* from the Orinoco fauna. With apparent colonizing influences from three different communities, each with endemic species of *Acanthobothrium, Potamotrygonocestus,* and *Rhinebothroides,* it is interesting that only a single species in each genus is established in the Maracaibo area. Perhaps this indicates something of the competitive abilities of the various parasites or perhaps it simply reflects uneven dispersal of suitable intermediate hosts. The latter seems more likely when Maracaibo is contrasted with the Orinoco (see below).

East Amazon. There is only one endemic species of stingray helminth found in this area, *Rhinebothroides freitasi.* One species, *Potamotrygonocestus amazonensis,* is a colonizer from the West Amazon fauna. Several species have been acquired secondarily from local teleostean hosts. The rest are derived from the Parana fauna through dispersal.

Orinoco. This area contains two different major faunas of potamotrygonid parasites, one derived through vicariance and one derived through dispersal from the Parana through the East Amazon with subsequent differentiation. To the first group belongs *Rhinebothroides venezuelensis, Potamotrygonocestus orinocoensis, Acanthobothrium regoi,* and *Echinocephalus daileyi.* To the second belongs *R. paratrygoni, Eutetrarhynchus araya, R. scorzai,* and *R. glandularis.* In contrast with Maracaibo, where a restricted fauna comprising a single species from three different faunas occurs, two faunas coexist in the Orinoco. If we include the occurrence of *P. amazonensis* from the West Amazon fauna via dispersal and the occurrence of *Megapriapus ungriai* from local teleosts, more than two faunas coexist. That this coexistence is real may be attested to by the fact that specimens of all three species of *Rhinebothroides* known from the Orinoco have been collected in a single host.

One difference between the Maracaibo and Orinoco faunas may be the occurrence of at least three species of potamotrygonids in the Orinoco and only one in Maracaibo. Brooks' (1981b) analysis of the host-parasite relationships of potamotrygonids using the analytical technique described herein placed two of those hosts, *Potamotrygon hystrix* and *P. reticulatus,* in different parts of the cladogram. *P. reticulatus* was related more closely to the Parana rays examined (*P. falkneri*), and *P. hystrix* was more closely related to *P. circularis* from the West Amazon. The Orinoco thus acts like two or three communities that have joined to form a larger community.

Two features suggest that the Parana–East Amazon–Orinoco connections are more recent than the Parana–West Amazon–Orinoco (and Maracaibo–Magdalena). First, *R. venezuelensis* occurs in Maracaibo (rather than *R. scorzai* or *R. glandularis*). Second, with the exception of *R. freitasi* and *R. glandularis,* none of the species representing the first pattern above has differentiated. Despite much dispersal subsequent to differentiation and some dispersal leading to differentiation, the historical constraints on the structure of parasitic associations of potamotrygonid stingrays are evident. Whether any of these conclusions are valid only for potamotrygonids and their parasites or for other species associated with these stingrays can be tested by future studies.

We have attempted to show that it is possible to establish the relative recency of co-occurrence among members of an association by using two types of data, phylogenetic history and biogeographical position. Species may co-occur because they evolved together or because at least one dispersed from its area of origin. In addition to their historical (H) or recent (R) geographical co-occurrence, species exhibiting ecological interactions, that is, those associates forming a community, may express in their interrelations traits inherited from their ancestors (historical [H] traits) or traits that evolved with them (recent [R] traits). In fig. 6.10, the four possible classes of causes for observed interactions are summarized. Since the community is an interactive unit, addition of one species due to colonization could lead to changes in another. For example, a colonizer with an R/H profile (colonizer, retains primitive life history traits) could, through

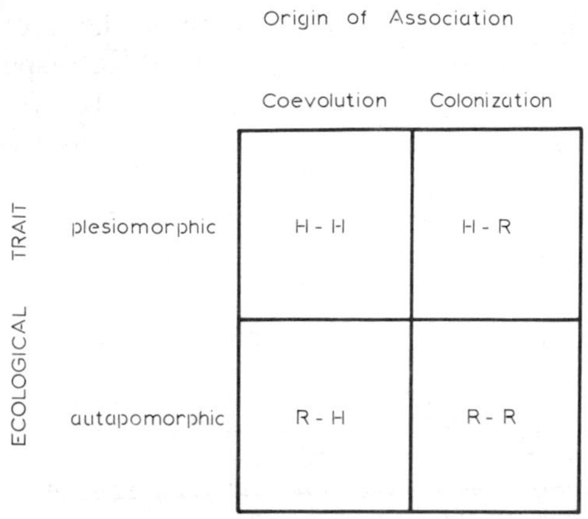

Origin of Association

		Coevolution	Colonization
	plesiomorphic	H - H	H - R
	autapomorphic	R - H	R - R

(Left axis label: ECOLOGICAL TRAIT)

Fig. 6.10. Two-by-two contingency table showing possible evolutionary explanations for observed ecological associates and individual ecological life history traits. H = historical, or inherited; R = recent, or modified, in the given case. (Redrawn from Brooks 1985.)

competition, cause an H/R shift in a local resident (no dispersal, shift in ecological life history trait). The question becomes, To what extent are the communities we observe the result of H/H histories? A second example drawn from work on rockfish parasites provides a clue as to how we can answer this question.

Rockfishes (*Sebastes*, family Scorpaenidae) comprise a large genus of mostly northern Pacific species. Holmes (1973) discovered that the digeneans *Psettarium sebastodorum* and *Aporocotyle macfarlani*, inhabiting the circulatory system of North American Pacific Coast rockfishes, exhibited little niche overlap. *P. sebastodorum* occurs almost exclusively in the heart, whereas *A. macfarlani* is almost entirely restricted to the blood vessels of the gill arches. Holmes assumed that parasite community structure is affected to a large degree by interspecific competition. From this, he concluded that the noninteractive site selection by the two species of blood flukes was de facto evidence of past competition.

Rohde (1979) examined the same data as Holmes. He assumed that natural selection could explain the observed differences in site selection easily. All one need do is invoke the notion that selection would favor flukes clumped together in the host because they would have a higher probability of mating. His conclusions: one need not invoke competition to explain the observations. The data are evidence of the effectiveness of natural selection.

	COEVOLUTION	COLONIZATION
Plesiomorphic	BROOKS (1980)	PRICE (1977,1980)
Autapomorphic	ROHDE (1979)	HOLMES (1973)

Fig. 6.11. Two-by-two contingency table from fig. 6.10 with names of authors who have supported different interpretations concerning the evolution of blood flukes and rockfishes along the western coast of Vancouver Island.

Price (1977, 1980) also examined Holmes' data. His assumption was that parasite communities, because they are generally found to be below MacArthur and Wilson (1967) equilibrium, are never established and persistent for long periods but are always young colonizing communities. He further asserted that generalized parasites are young colonizers and specialized parasites are not. Given the two specialized parasites in a young colonizing community, Price concluded that the parasites were preadapted to their specialized niches and therefore colonized them. Competition and selection played no role in their site selection in rockfish.

Finally, Brooks (1980) suggested that the parasites could be part of a coevolving fauna, that is, a vicariant ecological association, exhibiting plesiomorphic ecological life history traits for site specificity (habitat loyalty). The four different opinions each correspond to one of the general classes listed in fig. 6.10, and are shown in fig. 6.11. Brooks (1980) suggested that each of the four explanations predicted an explicit set of phylogenetic relationships and shifts in characters and/or hosts.

For Brooks' contention (the H/H association) to be correct, the two different parasites must exhibit a vicariant (= coevolving) relationship with the same host group and must exhibit plesiomorphic life history traits, as summarized in fig. 6.11, H/H. Price's hypothesis (fig. 6.11, H/R) would not differ too greatly, except that either *Psettarium* or *Aporocotyle,* or both, would be colonizers in rockfish, so the host-parasite relationships indicated by each parasite group would differ. Rohde's hypothesis (fig. 6.11, R/H) involves hosts and parasites that are together for a considerable period of time, during which natural selection promotes progressively narrower site selection. Finally, we consider Holmes' hypothesis of competitive interactions mediating site selection (fig. 6.11, R/R). Corroboration of this class of hypotheses requires both elements of dispersal and the expression of autapomorphic traits relevant to the dispersal.

Holmes and Price (1980) attempted to analyze the *Psettarium/ Aporocotyle*/rockfish system according to the above protocol. They concluded that there was no historically determined component in the interaction. Both species of parasites were deemed to be colonizers of rockfish and were "ecological specialists" in other hosts. Note that this corresponds to the situation found in fig. 6.11 (H/R), in which the ecological life history traits exhibited are plesiomorphic, that is, historically constrained. Thus, Holmes and Price were incorrect in asserting that their conclusions supported strict ecological determinism. They also asserted that ". . . the genealogies of parasites and their hosts are clearly not congruent (eliminating all 4 cases of Brooks' Figure 4)." Brooks' "figure 4" refers to fig. 6.11. Note that for only the H/H and H/R cases are host and parasite genealogies congruent. Note also that the H/R case corresponds to the conclusions Holmes and Price drew. Thus, they were also incorrect in saying that they had eliminated all four cases Brooks (1980) proposed. These inconsistencies in their conclusions led to a reexamination of Holmes and Price's phylogenetic hypotheses (Brooks, 1985). Regarding *Psettarium,* they wrote: "The genealogy of three species of *Psettarium* is not clear. . . . However, it does not appear that *P. japonicum* and *P. tropicum* are more closely related to each other than to *P. sebastodorum.* The male terminal genitalia of *P. sebastodorum* are more similar to those of *P. japonicum* than to those of *P. tropicum,* suggesting the genealogy of Figure 2B." Upon examining the primary literature, it was found that the similarities in male terminal genitalia of *P. sebastodorum* and *P. japonicum* are based on their possession of a cirrus sac (a bag containing the male intromittent organ)—*P. tropicum* lacks a cirrus sac. However, the presence of a cirrus sac is plesiomorphic not just for fish blood flukes, but for all flukes. Therefore, its absence in *P. tropicum* is not indicative that *P. sebastodorum* and *P. japonicum* are each other's closest relatives. On the other hand, *P. tropicum* and *P. japonicum* are both wide-bodied, whereas *P. sebastodorum* is slender and elongate. To determine which trait is plesiomorphic Brooks examined close relatives (outgroups) of *Psettarium* spp. Manter (1954) had stated that the differences between *Psettarium* and the genus *Cardicola* were "not clear" and "seemed rather inadequate." Holmes (1971) stated, "These two genera [*Psettarium* and *Cardicola*] are very similar, and certainly do not belong to different subfamilies." Of the nine species of *Cardicola,* some are elongate and some have wide bodies. Finding no clear-cut support within *Cardicola,* two other related species were examined—*Paracardicola hawaiiensis* and *Neoparacardicola nasonis.* Both of these species have two testes, the plesiomorphic condition for flukes. Members of *Cardicola* and *Psettarium* have only one testis; thus, they would seem to form a monophyletic group based on that trait. Both *P. hawaiiensis* and *N. nasonis* are elongate worms, suggesting that elongate bodies are plesiomorphic for the group. Thus, *P. tropicum* and *P. japonicum* would be closest relatives based on sharing the apomorphic trait of wide bodies. *Psettarium* is supposed to be discrete from *Cardicola* by virtue of possessing a bend in the posterior margin of the body at the point of the male genital pore. However, *N. nasonis* exhibits that trait, making it either a convergent trait or a plesiomorphic trait among these flukes. Thus, there is no real support for *Psettarium* being a monophyletic group distinct from *Cardicola. Cardicola,* on the other hand, has been distinguished only by virtue of the fact that it lacks the body bend of *Psettarium;* thus, there is no character support for *Cardicola* as a real group either. We will return to this point later.

The genealogy for *Aporocotyle* presented by Holmes and Price suffers from similar analytical flaws. Much of their cladogram is based on the presence or absence and arrangement of body spines and number of testes. All members of *Aporocotyle* have multiple testes, as do members of their closest relative, *Sanguinicola.* However, as mentioned before, the plesiomorphic number of testes in flukes is two. Holmes and Price's genealogy of *Aporocotyle* arranges species beginning with those having more than 130 testes and terminating with those having twenty-five to thirty-two testes. Outgroup comparisons would support a reversal of that trend. In addition, enlarged lateral body spines are used as indicators of relationships among species of *Aporocotyle,* even though *Sanguinicola* spp. have them as well. The phylogenetic hypothesis best supported by those characters reinterpreted in strictly phylogenetic terms is depicted in fig. 6.12.

In fig. 6.13, the cladogram for *Aporocotyle* and one for the three *Psettarium* species are displayed, with their host groups superimposed in place of species names. Although there is not much resolution, there is no evidence to support host-switching. The possible vicariant association between the two groups of blood flukes and

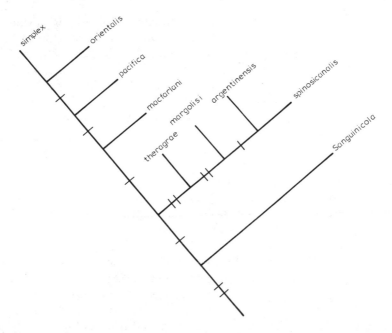

Fig. 6.12.   Cladogram depicting phylogenetic relationships of species of *Aporocotyle*, a genus of blood flukes inhabiting fish. (Redrawn from Brooks 1985.) Slash marks on branches represent synapomorphies uniting taxa.

the various host groups could be better examined if species of *Psettarium* occurred in cottoids or scombroids. Interestingly, four species of *Cardicola* occur in cottoids and three in scombroids. At least one of those occurring in cottoids, *C. laruei,* has a wide body like *P. japonicum* and *P. tropicum.* (The remaining two species of *Cardicola* inhabit mugiloid and labroid fishes, respectively.) Fig. 6.14 is fig. 6.13*a* redrawn to include the species of *Cardicola;* in fig. 6.15, the relative relationships of the fish groups involved and their associated blood flukes are shown.

The new analysis demonstrates no incongruence among the host and parasite cladograms. However, the vast majority of species of fish in each of the host groups have no blood flukes reported from them. This suggests to us that although both groups of blood flukes have coevolved with their hosts, that coevolution has been marked by independent failure of the two groups of flukes to persist in all evolving host lineages. The absence of parasites from many hosts does not indicate the degree of relationship among those surviving.

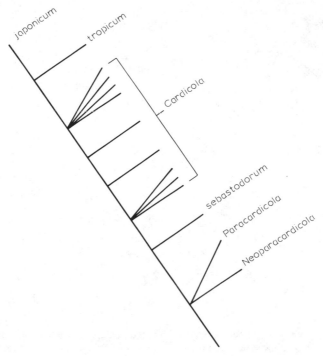

Fig. 6.13.   Phylogenetic relationships of *Psettarium* spp. (*P. japonicum, P. tropicum, P. sebastodorum*), *Cardicola* spp., and two outgroups (*Paracardicola* and *Neoparacardicola*). (Redrawn from Brooks 1985.)

It is not inconsistent with coevolution that in only one known case do members of *Aporocotyle* and *Psettarium-Cardicola* occur in the same host. This is especially true if the hosts in which the co-occurrence happens are relatively primitive, as they are in this case.

Having established the temporal context of the association between *A. macfarlani, P. sebastodorum,* and *Sebastes* spp., we may now examine the possibility of interactive or noninteractive factors determining the site selection exhibited by the parasites. For this we take the two parasite cladograms and superimpose infection site at the ends of the branches rather than the species names. We then estimate the ancestral conditions by means of "Farris optimization" (see Kluge and Farris 1969; Farris 1970) (fig. 6.16). Having those data, we may see that *A. macfarlani* occurs in the gills because its ancestor was a gill-inhabiting parasite (fig. 6.16*a*), but *P. sebastodorum* is most likely descended from a species that inhabited the

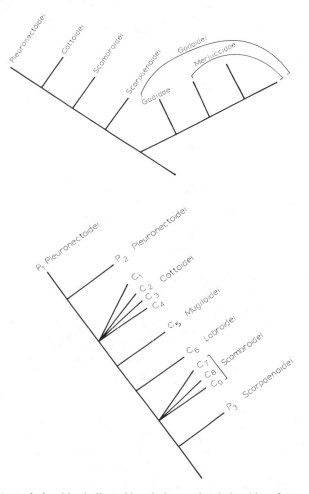

Fig. 6.14.   Host relationships indicated by phylogenetic relationships of *Aporocotyle* spp. (*top*). Host relationship indicated by phylogenetic relationships of *Psettarium/Cardicola* spp. (*bottom*). (Redrawn from Brooks 1985.)

mesenteric blood vessels (fig. 6.16*b*). Thus, *P. sebastodorum* exhibits an autapomorphic trait for site selection. This would appear to support Rohde's hypothesis, for that species at least. However, no other species in the group shows a similar tendency. On the other hand, the shift in traits would not seem to be explicable in terms of competitive exclusion, because *A. macfarlani* does not occur in the mesenteric blood vessels. The change in site by *P. sebastodorum* does not seem to be correlated with an extrinsic variable; so we

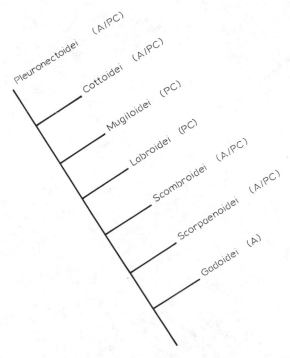

Fig. 6.15. Summary set of host relationships supported by both *Aporocotyle* spp. (*A*) and *Psettarium/Cardicola* spp. (*PC*). Note congruence. (Redrawn from Brooks 1985.)

conclude that the structure of this ecological association is histori-cally determined. This contrasts with the South American freshwater stingray parasite associations that include some colonizers.

We do not believe that the degree of historical structuring shown by parasite communities, such as the ones discussed above, is anom-alous. A highly analogous situation can be shown by comparing the roosting preferences of various species of cave-dwelling bats in Af-rica and in New Guinea (Hill and Smith 1984) (fig. 6.17).

If historical constraints play a major role in determining ecological structure, we would expect to find evidence of them at higher sys-tematic levels than the species. Brooks, Glen, and O'Grady (1985) recently presented a phylogenetic analysis of the sixty-three families of digeneans (to which *Psettarium, Cardicola,* and *Aporocotyle* be-long). Only 26% of the groupings and 35% of all branches on the phylogenetic tree were characterized by ecological changes. Classes of ecological traits considered were (1) changes in developmental

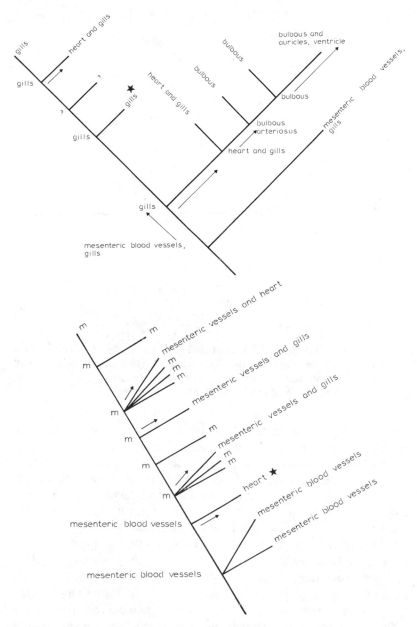

Fig. 6.16.   Summary of site-selection traits for species of fish blood flukes. *a, Apo-rocotyle*. Arrows indicate required evolutionary changes; question mark indicates unknown data; star indicates *A. macfarlani. b, Psettarium-Cardicola*. Arrows indicate required evolutionary changes; star indicates *P. sebastodorum*. (Redrawn from Brooks 1985.)

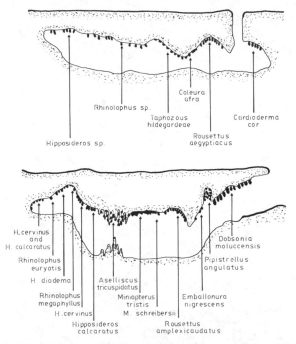

Fig. 6.17. Partitioning of roosting sites by different species of cave-dwelling bats in Tanzania (*top*) and on New Ireland Island. Note relative positions of species of *Hipposideros, Rhinolophus,* and *Rousettus* in each cave, suggesting a historical influence in site selection. (Modified from Hill and Smith 1984.)

programs producing increasing numbers of infective larvae, (2) changes in encystment behavior of infective larvae, (3) shifts in first host, (4) acquisition of and shifts in second hosts, (5) shifts in final host, and (6) changes in site of infection in the final host. Ecological changes definitely lagged behind morphological changes. Trends seen in species-level analyses are found in family-level analyses as well, attesting to the influence of historical constraints on ecological diversification. Although both the stingray and rockfish examples are based on host-parasite systems, we believe they represent generalizable associations. The methods used to analyze their structure are applicable to all organisms. Ross (1972a, 1972b) examined the phylogenies of several insect clades and concluded that ecological shifts accompanied speciation in only approximately one out of every thirty speciation events. A similar trend has been found among plants. Gilmartin (1983) concluded a study with this observation: "Thus, the [phylogenetic] trees, in particular the one for *Phytarrhiza* spe-

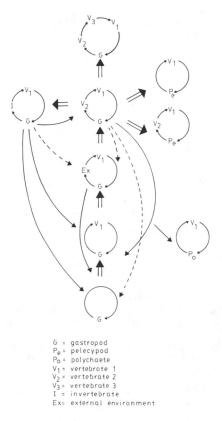

G = gastropod
Pe = pelecypod
Po = polychaete
V₁ = vertebrate 1
V₂ = vertebrate 2
V₃ = vertebrate 3
I = invertebrate
Ex = external environment

Fig. 6.18. Diagram summarizing evolutionary changes in life cycle patterns (host associations) for digenetic trematodes, beginning with the most plesiomorphic (*top*). Arrows with double lines indicate primary derived conditions, whereas arrows with single lines indicate secondary derived simplification (reversals). (Modified from Brooks, O'Grady, and Glen 1985.)

cies, are supported by character combinations that go far beyond the habit descriptors themselves.''

Furthermore, the extent to which ecological life history traits are historically constrained is the extent to which we should find that the evolutionary diversification of ecological life history traits in a group exhibit an irreducible historical component in all transformations. The phylogenetic transformations that have occurred among the life cycles of digeneans are depicted in fig. 6.18. Hosts have been added, replaced, and lost during the evolution of this group of parasites, but in all transitions at least one of the ancestral host types is retained. In no case does a life cycle evolve that has no historical component. In retrospect, such an orderly set of changes might appear somewhat like a periodic table of determinate life cycle types, but that is only because of the determinate nature of history. Note also that the number of changes in life cycle patterns lags considerably behind the number of species for which life cycles are known (about 500). This is an indication that some communities can be formed by historical constraints, which then become an additional source of constraints.

Measuring the Relative Entropy and Organization of Communities

In the absence of ecological interactions between species, the particular species and the numbers of organisms per species are determined by properties inherent in the ontogenetic programs and population biology of each species, as given in chapters 3 and 4. These lower level constraints form those initial conditions constraints on community structure that are not dependent on other species. They are thus the source of community-level emergent self-organization, or inherent information. This is analogous to the information provided by the DNA in a zygote at the ontogenetic level and that provided by variable ontogenies at the population level. The particular ecological interactions among species are also initial conditions constraints, because they represent finite energy-use capabilities determined by ontogenetic programs. However, such ecological interactions involve more than one organism and are in that sense partly external to the organism. They are the source of self-imposed organization at the community level, loosely analogous to cohesion at the population level and to epigenetic phenomena at the ontogenetic level.

The statistical entropy of a community may be expressed as a function of the relative numbers of organisms representing each species of organism and of their ecological interactions. This defines the energy flow pathways characteristic of the association. Entropy production may be manifested in changes in the relative numbers of organisms per species (which affects the relative amount of energy taken up by each species) or as changes in the topology of the energy flow relationships. Let us consider first the topologies alone. In fig. 6.19, four different associations represented as directed graphs indicating the direction of energy flows are depicted. The second graph from the top differs from the top graph by the addition of species D with its connections. The statistical entropy of the top graph is

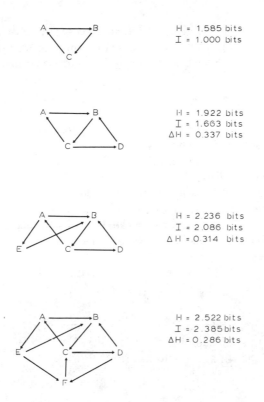

Fig. 6.19. Four food webs represented as directed graphs. Directions of arrows indicate direction of net energy flow resulting from interactions.

$$H_a = -(3)(1/3 \log_2 1/3)$$
$$= 1.585 \text{ bits}$$

and for the second graph is

$$H_b = -(3)(1/5 \log_2 1/5) - (2/5 \log_2 2/5)$$
$$= 1.921 \text{ bits}$$

Table 6.1 Comparison of Measures of Connectance (C), Organization (I), and Complexity (= statistical entropy) (H) for Four Ecological Associations Shown in Fig. 6.19. Rows proceed from top to bottom corresponding to the order in fig. 6.19

C	I	H
0.50	1.000	1.585
0.41	1.664	1.921
0.35	2.086	2.236
0.33	2.385	2.522

where the probabilities calculated are based on the number of flow
pathways emanating from each species in the assemblage. The entropy increase of the system resulting from the addition of species
D is

$$H_b - H_a = 0.436 \text{ bit}$$

The relative numbers of organisms of each species may also influence
the entropic behavior of the system and will contribute to the statistical entropy of the ensemble as well. We may produce a generalized formula for discerning the statistical entropy of a community
based on both components we have discussed.

Let X be the total number of energy flow connections among the
species present. Each species represents a probability class. Let x_i
be the number of energy flow connections emanating from the ith
species. Therefore

$$\sum \frac{x_i}{X} = 1$$

In addition, let N be the total number of organisms in the community
as estimated in a census of the community. Let n_i be the number of
organisms of the ith species found in N. Further

$$\sum \frac{n_i}{N} = 1$$

After normalizing the two probability terms in the same manner that
characters and tree topology were normalized in chapter 5, the statistical entropy of a given community becomes

$$H = -\sum \frac{x_i n_i}{XN} \log_2 \frac{x_i n_i}{XN}$$

A special case of this occurs when all ecological interactions are
considered equiprobable. In this case, all x_i are equal, and

$$H = -\sum \frac{n_i}{N} \log_2 \frac{n_i}{N}$$

which is a commonly used form of the Shannon-Wiener diversity
index. This relates our commonsense notions of "diversity" to entropy. We hasten to acknowledge Pielou's (1974) admonition that

$$\frac{n_i}{N} = P_i$$

implied by the above is strictly valid only for very large numbers. This is because the Shannon-Wiener formulation contains no inherent energy flow asymmetries and thus represents an equilibrium view. Equilibrium systems exhibit such periodic behavior that any measure of statistical entropy is strictly valid only in the realm of very large numbers, where any macroscopic trends can be detected. However, we have suggested that there are always inherent asymmetries in energy flow pathways in communities, rendering them nonequilibrium systems. Hollinger and Zenzen (1982) noted that nonequilibrium systems respond too quickly to disturbances for their dynamics to be governed totally by "the law of large numbers." Thus, we think biologically meaningful information can be obtained using our more general formulation. If ecological associations are nonequilibrium systems, their organization results from the dynamic interaction of entropic behavior of energy flows through the community and of the more conservative rate of change in ecological life history traits of the species the community comprises. We should find that the actual number of energy flow pathways possible for a given association will always lag behind the maximum possible. This type of dynamic process would produce community-level organization in a manner analogous to the production of information in our simple evolutionary dynamic (chapter 2) and the computer model (chapter 4). That is

$$I = \log A - H$$

where I is the information content or organization of the system, $\log A$ is the maximum topologically possible entropy of energy flow pathways, and H is the statistical entropy of the ensemble.

Pimm (1982) made use of a measure, which he called *connectance*, designed to provide an estimate of the internal organization of a given food web. Connectance is defined as

$$C = \frac{A}{B}$$

where C is the connectance, A is the number of actual links among the members of the community, and B is the maximum possible topological links. This can be equated with our notation thusly:

$$C = \frac{H}{\log_2 B}$$

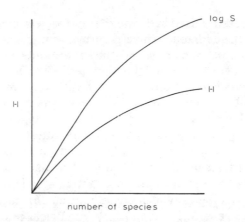

Fig. 6.20.   Schematic summary of findings by King and Pimm (1983). Two measures of community diversity were analyzed: log S, where S = the number of species in the community; and H, the Shannon-Wiener diversity index. As the number of species increases in real and simulated communities, H lags farther and farther behind log S.

where H is calculated solely on the basis of energy flow links. Pimm also noted that for real systems the number of species (S) increased as C decreased, such that

$$SC = k$$

where $k = 4$–5. May (1983) noted more generally that for large S

$$C = \frac{nS}{S^2}$$

where $SC = n$. Such an observed relationship would be found if each new species coming into an ecological association interacted ecologically with a relatively small number of the species already present. Using our notation, $\log_2 B$ would increase faster than H; thus I would increase and the association would be in a state of dynamic nonequilibrium. This state would be characterized by increasing complexity and self-organization without postulating any negentropic or destructive organizing force. King and Pimm (1983) provided empirical and theoretical evidence supporting the existence of such states (fig. 6.20).

In fig. 6.19, we have depicted four simple associations. Values of connectance (C), information content (I), and entropy (H) of the ensembles are presented in table 6.1. Note that as C decreases, both I and H increase.

The second measure used in community ecology is *evenness*. This is a measure of the degree to which populations of different species in an association approach or depart from equal numbers of members or equiprobable distribution of members. Evenness is defined as

$$E = \frac{H}{H_{max}}$$

where H is the Shannon-Wiener diversity value and H_{max} (or log A) is the equilibrium value, achieved if all populations are equal sized or equally distributed. Evenness is thus analogous to connectance except that it is based on relative population size rather than on energy flow interactions. Therefore, the same organization arguments stated for connectance hold true for evenness. Evenness values would be in nonequilibrium states if species occurred in highly asymmetrical population sizes, for example, if most species in communities were rare. If we subtract our connectance or evenness values from unity, the measures become estimates of *order* (Q; Landsberg 1984b) or redundancy (R; Gatlin 1972) in the community.

Some community ecologists have suggested that theories of community evolution need to answer the question of why there are so many rare species of organisms. One answer is that competition for energy sources within communities produces finer niche partitioning, leading to ever-greater energy flow partitioning; the more finely energy flows are partitioned, the smaller the population size that can be supported. In a top-down view of evolution, this functional subdivision of communities would lead to diversification of populations and speciation. Widespread species under this view are those which are not affected by competition to such an extent that they are modified evolutionarily.

We do not think evolutionary diversification is caused by energy flows or by competition for energy sources. Rather, we think that diversification results from partitioning of information sources, populations, and species. If this partitioning is due at least partly to geographical partitioning, we would expect, all other things being equal, that the population sizes of descendant species occurring in smaller (subdivided) areas than the ancestor would themselves be smaller as a consequence. Since geographical subdivision would affect many community members in the same way, we would expect a corresponding decrease in availability of some nutrient sources; thus, partitioning of information sources should constrain increases in descendant population sizes by decreasing some energy sources.

This would be especially true if, as we suggest, most speciation events do not involve ecological changes. Competition does not cause rarity of species but may act to constrain subsequent increases in population sizes. In our theory, rare species are not an anomaly, but widespread species are. We explain widespread species as those which (1) do not respond to geographical subdivision by speciating or (2) evolve the ability to use an energy source or sources that have not previously been used and have the ability to disperse from their place or origin. Thus, our explanation of the success of widespread species does not differ substantially from that suggested by most previous workers. However, we see widespread species as exceptions to the general tendency for species to be composed of more-or-less separate demes.

In addition, since energy flows are highly asymmetrical in communities and since energy transfers from one trophic level to another tend to be rather inefficient (although not so much that they are the determining capacity [Pimm 1982]), we would expect species occupying progressively higher trophic levels to be represented by progressively smaller numbers of individuals. This also is not a new idea, although we think that its integration into a larger framework is original.

According to our theory, the most highly organized community will be one in which the maximum possible value of I is achieved; thus, the higher the I value, the greater the organization. Furthermore, the extent to which the community structure is historically determined is the extent to which a given community is *self-organized*. Community organization increases as the actual entropy increases lag behind the maximum possible entropy increases, in accordance with the general dynamic presented in chapter 2 (fig. 6.21). Thus, organization and self-organization are closely related to minimum entropy increases at this level as well as lower levels.

States of minimum entropy increase are associated with high degrees of internal stability and resistance to extrinsic perturbations. Do communities tend to assume high stability, or minimum entropy, states, subject to proximal influences?

Pimm (1982) has discussed patterns of species interaction in food webs in detail. He has summarized the general attributes of stable food web topologies and the factors that contribute to these topologies (Pimm 1984). Twelve attributes were detected. Of these, two may be artifacts of the manner in which data are collected and one is "based on little evidence." The major factor is population dynamics; dynamic constraints predict eight of ten possible attributes.

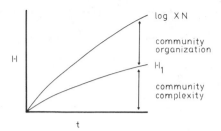

Fig. 6.21. Schematic representation of the relationship between complexity and organization in evolving communities. N = maximum number of species possible given number of organisms in community (each organism represents a different species); X = maximum number of possible food web links; H_c = entropy for the observed state of the community calculated on the basis of observed numbers of species, organisms per species, and food web links.

Simple biological realities are also important and predict constraints on five of the ten attributes. In addition, Pimm (1982) figured seven web topologies derived from one or more of these attributes and contrasted five of these topologies with topologies not usually observed in stable food webs. These attributes include:

1. There is an absence of "loops," predators without prey, and singular systems (fig. 6.22).

2. Webs are not too complex; connections between species decreases with increases in the number of species.

3. Food chains are short (three to four trophic levels, fig. 6.22a).

4. Omnivores are scarce (fig. 6.22b).

5. Omnivores, when they occur, usually feed on species in adjacent trophic levels (fig. 6.22c).

6. Webs dominated by insects and their predators and parasitoids form more complex patterns, which are exceptions to 4 and 5 above (fig. 6.22d).

7. Systems that are "donor controlled" display patterns like 6 above and also provide exceptions to 4 and 5 (fig. 6.22d). Donor-controlled systems are webs that include detritivores and scavengers that must await the death or imminent deaths of other organisms (the donors).

8-9. Webs are not usually compartmented, and when compartments occur, they are correlated directly with habitat divisions (fig. 6.22e).

10. Patterns of overlaps in the prey used by a set of predators can be expressed as overlapping intervals along a line more often than would be expected by chance alone (fig. 6.22f).

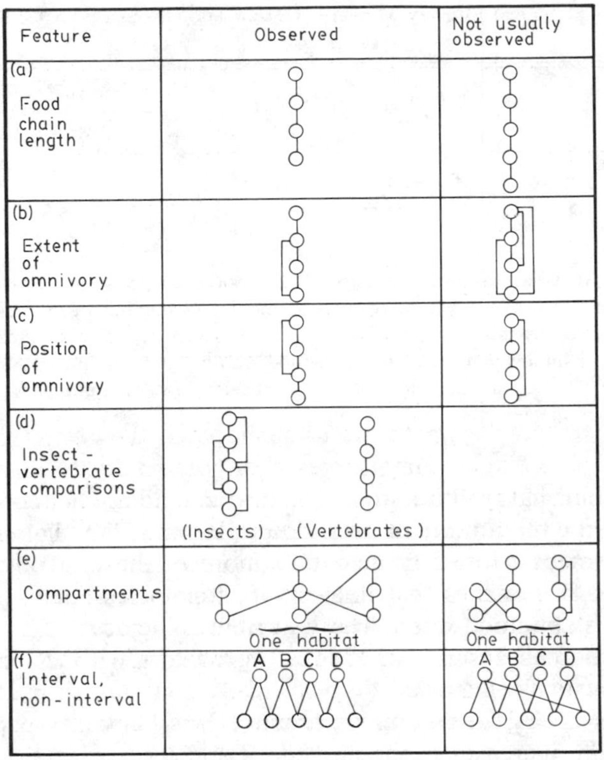

Fig. 6.22. Partial catalogue of features observed and not observed in real food webs. (Modified from Pimm 1982.)

Since food webs can be expressed as topological diagrams, we can estimate their statistical entropy. This would provide a measure of the organization of the web expressed in entropic terms. We calculated such measures for the food webs shown in fig. 6.22 and compared the measures of "observed" and "not usually observed" food web patterns. The results of such calculations for the topologies in fig. 6.22 are shown in table 6.2. Note that in each case the "observed" topology has a lower entropy configuration than the "not usually observed" topology. The extent of omnivory when webs include insects, scavengers, and detritivores also results in a higher entropy configuration than that in "simpler" vertebrate systems.

These results present an interesting correlation. Low entropy values would seem correlated with observed food web patterns, with exceptions (i.e., insect omnivory) being explainable by proximal biological attributes. Perhaps there is an underlying entropy principle

Table 6.2 Statistical Entropy Measures (H) for Food Web Topologies Shown in Fig. 6.21

Feature	$H = -\Sigma\,(p_i \log_2 p_i)$	
	Observed	Not usually observed
a. Food chain length	1.919	2.522
b. Extent of omnivory	1.906	2.000
c. Position of omnivory	1.906	2.000
d. Vertebrates (and large invertebrates)	1.919	—
Insects, detritivores, and scavengers	2.271	—
e. Compartments	2.875	$2.281 + 1.584 = 3.865$
f. Interval, noninterval	3.125	3.128

at work that makes both the dynamics and biology of food web patterns predictable. Pimm noted that community resilience is directly affected by food web structure and that resilience is also correlated with rapidity of nutrient cycling and loss. The more rapidly a nutrient is lost from a community, the more resilient the community is in terms of that nutrient. The causal question is this: Is food web structure determined by nutrients, that is, the "flux" of energy sources forming part of the boundary conditions, or is food web structure determined by historically constrained evolutionary events involving the various members of the community and affected proximally by boundary conditions? Our theory asserts the latter.

The faster a nutrient is lost (dissipated) from a community, the smaller should be the entropy production associated with the cycling of that nutrient. This provides a direct connection between community resilience and the strictly thermodynamic conception of the principle of minimum entropy production. We have already seen that those communities observed are also minimum entropy configurations in terms of their food web structure. Thus, just as with ontogeny, in the evolution of ecological communities we can see that the principle of minimum entropy production is a highly general phenomenon. We would assert that any given community will assume the most resilient configuration of nutrient cycling possible, given the initial constraints of the inherent food web structure determined by the evolved functional characteristics of the component species and will fluctuate around that most resilient structure if amount of available nutrients (boundary conditions) fluctuate.

In our theory the relationship between community stability and community resilience is as follows: stability is indicated by the degree of self-organization, that is, the macroscopic information con-

Table 6.3 Summary of Causes of Food Web Patterns

Pattern number		Is pattern uncertain?	Is pattern explained by:		
			Dynamics?	Biology?	Energy?
i	No loops, predators without prey, singular systems		Yes	Yes	Yes
ii	Limits on complexity		Yes	Yes	
iii	Food chain lengths are limited		Yes		Perhaps
iv	Omnivores are scarce		Yes		
v	Omnivores feed on adjacent trophic levels		Yes		
vi	Insects and their parasitoids are exceptional to patterns 4 and 5		Yes		
vii	Scavengers feed on many trophic levels	Yes	Yes	Yes	
viii	Compartments correspond to habitat divisions			Yes	
ix	Webs are not compartmented within habitats		Yes	Yes	
x	More predators than prey	Yes			
xi	The more species of prey a species exploits, the fewer species of predators it suffers	Yes			Yes
xii	Prey overlaps are interval		In part (from vi)	In part (from viii)	

SOURCE: From Pimm 1982.

tent or organization value for a given community, whereas resilience is measured in terms of minimum entropy production in nutrient cycling. Thus, stability stems from minimum entropy configurations of food web structure, and resilience stems from minimum entropy production nutrient cycling given the food web structure. We need not postulate that the nutrients themselves "cause" food web structure, by some mechanism as yet unknown. Both the historical/ structural and the energetic/functional considerations are reconciled by our theory.

A test of this postulate of our theory concerning the primary causality of community structure would be determination of the degree to which energy considerations explain observed food web structure for those examples we have cited. Table 6.3 is a reproduction of Pimm's table 10.1. It depicts twelve patterns, three of which are listed as "uncertain." One of these twelve patterns, one or possibly two of the certain ones, are *partially* explained by energy. Only one pattern is totally explained by energy and that is an uncertain pattern. All the rest are explained *totally* by dynamics and/ or biology. Thus, available evidence seems to support our view.

If our theory adequately explains the evolution of communities, we would expect to find a decreasing rate of increase in entropy (smooth or fluctuating) and an increase in organization through time for any given community. Rearrangements in interaction topology should diminish, and population sizes should assume relatively stable sizes (i.e., the variance of evenness estimates should decrease). For communities, there are two interrelated general classes of processes affecting entropy. The first is immigration and emigration; the second is the evolution of life history traits by the members of the ecological association. The effects of both processes may be additive or nonadditive. For example, new species coming into an area may add to the topology of the local ecological association without altering preexisting connections; a similar condition may obtain for emigration or extinction. A stable-state equilibrium point, or state of no net entropy change, for additive immigration/emigration extinction phenomena is achieved whenever immigration/emigration rates are equal (symmetrical), and there is no net change in topology of associations. At the level of individual species' life histories, an analogous process operates. It is the use of alternate equivalent life history pathways. For example, consider a parasite that can successfully utilize either of two different host species. We may measure the relative degree to which each host is used over time, predicting that an equilibrium point would be achieved, characterized by as great a degree of symmetry in host use as possible. Once that level has been achieved, departures from it will also act like temporary fluctuations around an equilibrium point, producing no net change in entropy.

Now consider additive changes that irreversibly affect the entropy. These involve the addition or deletion of interactions without changing the rest of the community topology. For example, if the parasite species considered above evolves in such a way that it utilizes a third host or if a third utilizable host immigrates into the community, establishment of a new ecological interaction produces an asymmetry, which should lead to an increase in entropy. Most likely, this would be reflected by an increase in the overall population size of the parasite as it expands in utilizing the new host. However, we would predict that over time the new three-host community would reach some stable state of relative symmetry. Cases of deletion of an interaction, such as removal of one of those host species, that do not otherwise affect the community topology would also cause an increase in entropy as population sizes of other members increase.

Nonadditive changes in communities are those which disrupt at least part of the preexisting topology of the community. Whereas the changes discussed previously were analogous to changes in structural genes, nonadditive changes in communities are analogous to changes in regulatory gene architecture.

Any such changes, either by immigration or evolution, would be successful only to the extent that the species already established could compensate with alternate ecological interactions. Therefore, we would expect their likelihood to decrease as the topology of a community became more complex. We might even expect that ecological innovations in speciose, complex communities would be rare, whereas those in species-poor communities would be relatively common. Recently, Jablonski et al. (1983) have documented such a pattern for evolution of phanerozoic shelf communities. Alternatively, any such change that was successful would affect the dynamics of the entire system. This would tend to produce an increase in entropy, driving the system away from an overall steady state. This period might initially be characterized by "ecological chaos" as old interactions were lost and new ones formed, new species or species traits added, and, perhaps, old species lost. However, as Hollinger and Zenzen (1982) pointed out, such a "forced departure from equilibrium" would be characterized by increasing reproducibility (= organization) of the system. This would occur as the new ecological interaction topology predominated and the old topology disappeared. Once that was accomplished, the entire system would exhibit a progressive decrease in rate of entropy increase, rising to a new steady state in a manner very much like that described by Johnson (1981). The new steady state would be characterized, like the equilibrium states described by Hollinger and Zenzen (1982), by stochastic (or chaotic) behavior within the initial conditions of the association. Boucot (1983) has suggested that this "punctuated" pattern of ecological change characterizes the paleontological record for the evolution of many communities.

Raup and Sepkoski (1984) examined a corollary of such widespread changes in community structure, namely, periodic widespread extinctions. They noted twelve major extinctions during the past 250 million years and suggested that they might be due to catastrophic environmental events. If such events are associated with the substantial changes in ecological relationships among species noted by Boucot, we have an indicator of the magnitude of change in boundary conditions required to substantially alter the functional relationships among species. Or, in other words, it is an indication of the magnitude of environmental change required for environmental selection to effectively cause an adaptive radiation. In the absence of changes in boundary conditions of such magnitudes, we would expect ecological traits to be conservative relative to morphological traits.

Historical Ecology and Competition

Community ecologists are currently embroiled in a debate about the role of competition in determining the evolution and organization of communities (see Lewin 1983 for a review of the controversy). Competition in this sense may be broadly construed as the co-occurrence of two or more species that require the same resource, which itself occurs in limited supply. One view asserts that competitive interactions should be tested statistically against a null hypothesis of random co-occurrence of species; another suggests that one should assume competitive interactions unless compelled to another explanation.

By asserting that there will always be an irreducible historical component to the explanation of the evolution of any community, we imply a different experimental framework. Specifically, the "null hypothesis" against which hypotheses concerning the evolutionary context of any particular ecological association should be tested is *neither* a "competition" nor a "random" model. Rather, it should be a historical model.

Consider the evolution of three associated clades (fig. 6.23). We will investigate five different models of the evolution of this association: (1) historical ordering, or coevolution in our sense of the word (fig. 6.23a); (2) competitive exclusion (*sensu* Holmes in the example given in this chapter) (fig. 6.23b); (3) random change in precursor ecological traits (fig. 6.23c); (4) unlimited polymorphism for ecological interactions (fig. 6.23d); and (5) competitive exclusion without historical constraints (fig. 6.23e). There are three different ecological life history traits expressed, a, b, and c, in fig. 6.23. Maximum entropy occurrence of those traits would be log 3, or 1.585 bits, since $H_{max} = \log A$, and $A = 3$ in this case. The minimum entropy occurrence of those traits would be the de novo appearance of each trait in one species. Probabilities for such an occurrence would have a denominator of 63; 3, for each ecological life history

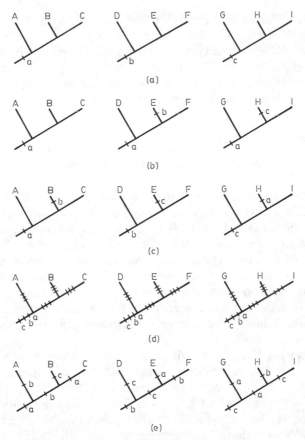

Fig. 6.23. Different historical contexts leading to contemporary ecological associations comprising species in three different clades with ecological life history traits
a, b, and c. *a,* All life history traits plesiomorphic (historical determinism). *b,* Life
history traits of E and H resulting from competitive exclusion, all ancestors having
trait a. *c,* Life history traits of B, E, and H resulting from competitive exclusion, all
ancestors having different traits. *d,* Unlimited polymorphism model, in which each
species has potential for all three ecological life history traits and observed expressions are result of nonhistorical proximal causes. *e,* Life history traits resulting from
random historical changes in all clades.

trait, times 3, for each clade, times 7, for the topology of a three-
taxon bifurcating tree (see chapter 4). Thus

$$H_{min} = -(3)(1/63 \log_2 1/63)$$
$$= 0.095 \text{ bit}$$

However, this H_{min} does not correspond to an evolutionary hypothesis, since de novo production of ecological life history traits under this model would assume that the ancestral species had no ecology. Therefore, a more reasonable H_{min} would be one that postulated the minimal entropy configuration compatible with an evolutionary hypothesis, as shown in fig. 6.23a, where

$$H_{min} = -(3)(2/63 \log_2 2/63)$$
$$= 0.474 \text{ bit}$$

The p_i value is 2/63 because nonterminal branches have a topological value of 2. We now have an H_{min} and an H_{max}, or log A, for the community. This means we can also calculate the maximum possible organization of the community, as information (I), in which

$$I_{max} = \log A - H_{min}$$

In the case of fig. 6.23a,

$$I = 1.585 - 0.474$$
$$= 1.111 \text{ bits}$$

Now let us consider entropy (complexity) and organization of the community under the influence of factors other than historical ordering. In fig. 6.23b, B, E, and H have come together through dispersal from different ancestral communities where they all exhibited ecological life history trait a. Subsequent to their current association, species E and H developed life history traits b and c, respectively, as the result of competitive interactions. The effect this has on the structure of the community is as follows:

$$H_b = -(2)(1/63 \log_2 1(63) - (6/63 \log_2 6/63)$$
$$= 0.513 \text{ bit}$$

where 1/63 is the p_i value for traits b and c, and 6/63 is the value for trait a, which occurs in a nonterminal position in all three lineages (thus $3 \times 2/63 = 6/63$), and

$$I = 1.585 - 0.513$$
$$= 1.072 \text{ bits}$$

Thus, we find that the community in fig. 6.23b is more complex (higher entropy) than that in fig. 6.23a, but that it is also less organized, since I is 1.111 bits in fig. 6.23a and only 1.072 bits in fig. 6.23b. The change is slight, but discernible. For fig. 6.23c, the three species in the community all exhibit random fixed changes in their ecological life history traits (this could also be achieved through competition). In this case

$$H_c = -(3)(3/63 \log_2 3/63)$$
$$= 0.627 \text{ bit}$$

and

$$I = 1.585 - 0.627$$
$$= 0.958 \text{ bit}$$

Once again, entropy increases and organization decreases relative to historical ordering. In fig. 6.23b,c, there is still some degree of historical constraint on the evolution of new ecological life history traits. The next two models are designed to avoid historical constraints.

First, fig. 6.23d represents a model we call the *unlimited polymorphism model*. That is to say, every species always has the ability to exhibit ecological traits a, b, and c, and the observed mix is due purely to the competitive interactions seen at any one time. For this model

$$H_d = -(3)(21/63 \log_2 21/63)$$
$$= 1.585 \text{ bits}$$

Notice that $H_d = \log A$. Thus

$$I = 1.585 - 1.585$$
$$= 0$$

There is no organization inherent in the community. It is a maximum entropy, minimum organization configuration. Now consider the model represented in fig. 6.23e. In this case, traits a, b, and c are fixed randomly among the members of the clades, and no necessity for competition is required (even though it might have occurred, producing the same results). Here

$$H_e = -(3)(7/63 \log_2 7/63)$$
$$= 1.057 \text{ bits}$$

and

$$I = 1.585 - 1.057$$
$$= 0.528 \text{ bit}$$

While this model is not a maximum entropy model, it is nonetheless a high entropy, low organization phenomenon.

From the preceding simple examples, we may observe several general trends. First, the lowest entropy, maximum organization model is one in which all ecological interactions are historically determined. Second, models of competition, and/or random change that are historically constrained show increased entropy and decreased organization. Third, models of competition and of random change without historical constraint are very high entropy and low organization phenomena. Maximal entropy is attained by a model that assumes no intrinsic constraints, or self-organizing capabilities, at all.

We conclude that in the absence of extrinsic perturbations, communities, like developing organisms and evolving populations and species, will exhibit an evolutionary tendency toward increasing self-organization (i.e., historical constraint). That is, increases in entropy signaled by the evolution of new ecological life history traits will be minimized. The effects of extrinsic perturbations will be disorganizing rather than organizing. However, unless all historical constraints are overcome in all cases, the disorganizing effects of the extrinsic factors will not produce disorganized ecological associations. This historical component is the source of stability and resilience in ecological associations; the greater the I value, the more perturbation an ecological association can withstand without becoming disorganized, and the greater will be the tendency to maintain or revert to a historically determined organization in the face of or following perturbing influences.

Summary

The evolution of organisms that feed on other organisms coupled two types of entropic phenomena: (1) the evolution of biological information specifying ecological life history traits and (2) the flow of energy through energy-transfer systems. The coupled entropy flows operate at different rates with new information specifying energy flow pathways lagging behind the energy flows themselves. Resistance to equipartitioning of energy occurs, leading to the formation of stable nonequilibrium ensembles of different species. The result is the emergence of a functional level, that of ecological communities, higher than that of the individual deme or species.

Two major predictions emerge from this view: (1) in a phylogenetic analysis of any clade, evolution of ecological life history traits will lag behind morphological evolution and (2) there will always be an irreducible historical component determining the structure of any ecological association. There is an experimental protocol available for testing these assertions, and we have provided some examples of its application at different levels of analysis. Available neontological and paleontological data seem to support our contention.

Proximal factors, both biotic and abiotic, may influence the behavior of any given ecological association. Instances of such perturbations may reveal previously unexpressed polymorphisms in ecological life history traits, or they may produce permanent changes in ecological life history traits through evolution of the species affected or through changes in the species composition of the association. In contrast to some previous ecological theories, our theory predicts that such external perturbations will decrease the organization of the association.

Two related methodological principles guide investigations into ecological associations under our theory. The first is that any causal explanation involving ecological associations must contain information about the history of the association in question. The second is that one must study more than one (in fact, at least three) ecological associations at a time. This is because hypotheses of historical context are comparative. Much of this chapter has been devoted to presenting protocols for accomplishing those goals. Although improvements are necessary, techniques are currently available to perform such investigations and, to a limited extent, such studies have already been done. Nonetheless, some data have been collected concerning ecological associations and evolution, including those in paleontology (see Ross 1972a, 1972b; Boucot 1975, 1978, 1982), and the following four trends present themselves: (1) ecological changes do not precede or accompany speciation very often; (2) speciation does not result in ecological changes very often; (3) species differences are rarely correlated with any independent ecological variable; and (4) ecological determinism has very little empirical support as a doctrine for evolution of ecological associations when evolutionary history is included in explanations.

Let us now consider ecological communities in light of our four basic principles of evolution. When considering the principle of irreversibility, we interpret ecological communities in the same way as any other component of the living world. This means they are nonequilibrium systems that are open to energy flows but closed or partially closed in terms of information. Thus, in terms of the principle of irreversibility, we find it possible to show a smooth reduction from highest levels to lowest levels of organization in biology. That this can be done is not so clear with respect to the principle of individuality.

If ecological interactions are causes of evolutionary change, knowledge of the evolution of ecological associations will be derived from observation of ecological interactions, which, as causal evolutionary units, will be found empirically to be independent entities. Species, as effects of the interactions, will be class constructs. We do not think the former assertion has been documented, nor do we think species are classes. On the other hand, if the evolution of species by means of entropic changes in cohesion and information causes changes in ecological interactions, knowledge of the evolution of ecological associations will be derived from observations of the evolution of life history traits and their ecological manifestations. Ecological interactions, as effects of evolution, will be class constructs. Species, as causal evolutionary units, will be found empirically to be independent entities. We believe this latter view is supported.

To illustrate this point, consider the simplest case, an ecological association involving two species, A and B (an autotroph and a heterotroph, host and parasite, predator and prey, etc.). We may substitute any other species for A and for B, but the topology of the interaction will remain the same. Thus, the interaction is a class construct. The only change in information content occurs when one or both of the species involved in the association changes. The species are the independent causal entities and the interaction is the effect, or the emergent property, of the species' evolution. Given an autotroph, herbivore and carnivore, we can predict the ecological association that will emerge.

Ecological associations are not individuals in the same sense that species and organisms are. Because ecological associations are a class of effects of the coevolution of species on this planet, one might think they have an ontological relationship with natural groups or monophyletic groups of species. Wiley (1980a) suggested we call monophyletic groups *historical groups,* because the connections among their differentiated members are historical rather than reproductive. Historical groups are products of evolution, just as ecological associations are. The observation of concomitant vicariant speciation by different individual species making up a set of ecological associations is not necessarily indicative that they are parts of a higher level of individuality, but rather that they are all affected by a law or laws of higher generality. Two other observations about the composition of ecological associations bear on this point. First, when a given apomorphic trait occurs independently in more than one member of a monophyletic group, it is explained as a parallel or convergent product of two different ancestors. But if members of the same species occur in more than one ecological association, it is not because the same species evolved twice, in each of the associations. Rather, members of the species colonized one of the associations after evolving in the other. Second, when two different traits evolve in a species, we speak of a polymorphic species. But if sympatric speciation occurs, an ecological association may include two separate species that are each other's closest relatives. Because historical groups are not strictly individuals, the occurrence of the two species cannot really be called a "polymorphism." Thus, the "individuality" of ecological associations as dissipative structures cannot be reduced smoothly through various levels of individuality in living systems. We conclude that ecological associations are class constructs in spite of their historical cores.

The principle of intrinsic constraints comprises two components when applied to ecological associations. The first is the sum of the intrinsic constraints on realization of new traits in all the species involved. The second is the set of constraints on the realization of new ecological life history traits, by means of evolution or colonization, within the association provided by the established members of the association. Simply put, this amounts to the rather trivial assertion that a new trait, or a species with a new trait, will be less likely to survive in a given ecological association if it is redundant with traits already possessed by members of the association. Alternatively, if that new trait or species is to survive, it will do so at the expense of an established species. In other contexts, this truism has been accorded some degree of importance as a "principle of competitive exclusion," and we do not wish to deny that it may occur. However, we do wish to point out that recognition of an inherent bias against redundancy in evolving systems is not evidence for an active process of competition in the Darwinian or neo-Darwinian sense. We are surprised to find redundancy in ecological life history traits at all, because that leads to competition that is disruptive of established ecological order. A property of dissipative structures is that they move away from conditions of high perturbation to stable states characterized by the minimum amount of fluctuation possible and tend to resist new perturbations to a great degree after that. Rather than being "powered" by competitive interactions, ecological associations as dissipative structures tend to minimize them.

This is not to say that such perturbations do not occur or are not important. The behavior of ecological associations under such conditions is considered an aspect of compensatory changes.

If the evolution of ecological life history traits exhibited by members of an ecological association is historically constrained, such traits should be analyzable just like any other trait. We have shown two cases in which that has been done, and the results support our contentions. We might also expect to find paleontological evidence that ecological life history traits evolve no more quickly than any other single kind of character and, to the extent that they are influenced by two levels of intrinsic constraints (those of each species' information system and those of the ecological association of which each species is a part), we might expect them to evolve even more slowly. Boucot (1982) has assembled an exhaustive account of paleontological evidence bearing on this issue. He concluded that ecological life history traits appear rather suddenly in the fossil record and persist unchanged for extremely long periods of time in lineages that may exhibit high rates of speciation nonetheless. He found no evidence of correlation between speciation per se and ecological shifts. Boucot accounted for this pattern of evolution of ecological traits by invoking Simpson's (1953) idea of "quantum evolution." But quantum evolution, or the sudden shift from one adaptive zone to a new one followed by adaptive radiation in the adaptive zone, was formulated as an explanation for transspecific evolution, a phenomenon we contend does not occur. Boucot's data are expected findings in our theoretical framework rather than the special cases that require special explanation under the doctrine of quantum evolution. By contrast, his studies show very little support for the notion that ecological, or functional, changes have anything causally to do with most speciation. Evolutionary biologists who wish to explain Boucot's data without reference to transspecific evolution may propose, ad hoc, that the fossil data do not adequately represent the evolutionary process. But consider the more than forty-member species of the genus *Haematoloechus,* a group of digenetic trematodes. Virtually all of them live in the lungs of frogs of the genus *Rana* and have life cycles involving aquatic pulmonate snails and dragonfly larvae as intermediate hosts. Their geographical distribution is worldwide, suggesting an evolutionary history stretching back to the Mesozoic. The species are morphologically distinct but ecologically virtually identical. We suggest that this is the rule rather than the exception. Ross (1972a) estimated that ecological shifts occurred approximately once in every thirty speciation events in various insect groups.

One outcome of seeing that functional changes lag behind structural changes in evolution is that food web topology rather than population fluxes should be primarily important in determining community structure. Another is the recognition that evolution produces species which, in the context of the next higher functional level—that of communities—are usually functional equivalents of their ancestors but may on occasion be functionally different. In a functionalist view of biology, structural novelties might be given a general name to denote their presumed roles in maintaining functional systems. Gould and Vrba (1982) have used the term *aptation,* subdivided into *exaptations,* or structural novelties having no functional significance when produced but later coopted by natural selection for a particular purpose, and *adaptations,* or structural novelties evolving in response to the functional requirements of any given selection regime. With our structuralist view, such distinctions are unnecessary and inappropriate descriptors. Functions are emergent properties of structure, and different structures may function in the same or different ways. We find empirically that different structures perform in the same way more often than they perform differently during the course of evolution.

The principle of compensatory change may well explain such ecological phenomena as biotic succession and ecosystem stability and resilience, the latter two as measures of the pool of potential information among species for ecological traits. We find evidence of compensatory responses in many such systems. For example, many plants grow at a faster rate after being cropped by a herbivore than before. This is not an adaptation to herbivory by the plant, but rather a predicted physiological response of an organism that has been perturbed from its steady state. When an adult tapeworm attaches to the intestinal lining of its vertebrate host, there is often in response a thickening of the intestinal epithelium at the site of attachment. Again, this is not an adaptation to defend against parasites, but a predictable response by an inducible tissue. What is common to both cases is that the *result,* not the cause, of such phenomena is conditions conducive to the survival of both. We would not expect to see any cases in which the opposite were true simply because they would be eliminated. Alternatively, we find evidence of compensatory changes in ecological associations involving overlap of energy or space requirements among species. For example, if two herbivore species occur in a given area, one capable of subsisting on rice or bamboo and the other capable of subsisting only on bamboo, we might expect to find the first species eating rice and the second

Table 6.4 Formulation of Several General Concepts Used to Describe Important Aspects of Community Structure in Terms of Statistical Entropy Measures

Term(s)	Formulation
Diversity (= complexity, = entropy)	$H = -\sum \dfrac{x_i n_i}{xn} \log_2 \dfrac{x_i n_i}{xn}$
Stability (= organization, = information)	$I = \log_2 (xn) - H$
Order (= redundancy)	$Q = 1 - \dfrac{H}{\log_2 (xn)}$
Connectance (= disorder of food webs)	$C = \dfrac{-\sum x_i \log_2 x_i}{\log_2 x}$
Food web order	$Q_C = 1 - C$
Evenness (= species' composition disorder)	$E = \dfrac{-\sum n_i \log_2 n_i}{\log_2 n}$
Species' composition order	$Q_E = 1 - E$

NOTE: x_i = Probability associated with ith ecological connection in a food web of x maximum possible food web connections; n_i = probability associated with proportion of organisms representing ith species in a sample of n organisms sampled in a community.

bamboo. This we would call a compensatory change of the general kind called *character displacement* (Brown and Wilson 1956). Both species occur in a given area, one capable of subsisting on rice or bamboo and the other capable of subsisting only on bamboo, we might expect to find the first species eating rice and the second bamboo. This we would call a compensatory change of the general kind called *character displacement* (Brown and Wilson 1956). Both species are able to survive because one is flexible in its requirements. If a third herbivore requiring rice arrived on the scene, and rice was in short supply, *competition* in our sense of the word (and, we think, in Darwin's sense) would occur—all three species could not continue to exist, at least not at initial population levels. We consider compensatory changes via interactions of flexible information systems with each other or with abiotic factors, embodied in the term *co-accommodation* (Brooks 1979), to be responsible for fine-tuning ecological associations.

A more general term for these phenomena is *density compensation*. Faeth (1984) listed nineteen studies of density compensation and showed that the effect is not due to insularity (or compensation due to boundary conditions constraints) and that it occurs only to

288   HISTORICAL ECOLOGY

the extent allowed by the inherent capacities of the expanding species. That is, the elimination of one species from a community will
result in a compensatory phenomenon only to the extent that some
other species' initial conditions allow it to utilize resources no longer
used by the removed species. There is no particular causal requirement that such unused energy sources be used.

The origin and evolution of ecological associations can be shown
to be a nonequilibrium entropic phenomenon. The interaction of
energy flows through ecological associations with relatively conservative evolutionary changes in ecological life history traits leads to
increase in organized complexity of ecological associations. Thus,
our theory predicts an ecological paradigm of coevolution. This view
leads to a unified characterization of a number of general terms used
to describe important aspects of community structure (table 6.4).
Furthermore, ecological associations can be characterized by three
of the four principles of evolution we have set forth. Some problems
are encountered when extending the principle of individuality to
these associations. They are clearly individuated physical systems
at any given time, but new associations can be constituted out of
already existing species. This means that they do not exhibit a higher
level individuality; ecosystems are not super-organisms even though
they have historical cores.

· 7 ·

Reprise
and Prelude

We have been primarily interested in accomplishing two goals. First, we wanted to provide a unified theory that shows how processes operating at different functional levels in biology all contribute to the evolved diversity that we encounter. This means we are striving for an evolutionary theory that does not rely solely on environmental determinism for evolution's direction and ordering. Our assertion that selectionism is an incomplete explanation of evolution does not amount to an attempted justification of orthogenesis as it is commonly understood. We think that such a characterization would be superficial. Orthogeneticists traditionally have postulated that population-level phenomena are irrelevant to evolution, which is seen as strictly a developmental phenomenon. Neo-Darwinism, as a reaction to orthogenesis, has traditionally asserted just the opposite (see Bowler 1983). Since both levels of phenomena occur, it seems likely to us that a robust evolutionary theory would incorporate both rather than excluding one or the other. To accomplish such a unification, we first need a common ontology, provided by the principle of individuality pioneered by Ghiselin (1974). We also need a common theme, which we found in history and irreversibility.

Second, in attempting to unify evolutionary theory, we needed to provide a common background dynamic and a common quantitative currency to show that the various functional levels could be linked as different manifestations of a more general phenomenon. To accomplish this, we have attempted to find an appropriate link between biological evolution and more general natural laws. That is, we have attempted to address the question of ultimate cause, or axiomatic behavior, as well as proximal cause in biological evolution. In this effort, we are not the first scientists to pose the question. We are not the first scientists to seek that link in thermodynamics and related concepts; in fact, we are not even the first to decide that nonequilibrium thermodynamics might hold the key.

To us, biological evolution is first and foremost a time-dependent, or irreversible, process. It is characterized as a process by the emergence of new ordered states from old ordered states. Systems exhibiting such behavior in general are described by statistical mechanics and nonequilibrium thermodynamics. If evolution is a process characterized by the production and maintenance of order due to extrinsic factors, that is, natural selection narrowly construed, there is an irreducible Maxwellian demon in the dynamics of the process that needs to be reconciled with the physics of time-dependent change. Nonequilibrium thermodynamics provides the theoretical framework for such occurrences as improbable but statistically possible states. However, if evolution is characterized by orderly changes inherent in the dynamics of organic processes, such as ontogeny and reproduction, we must show that no Maxwellian demon is necessary. Doing this requires that we demonstrate an increase in entropy accompanying biological processes. Entropy is usually measured in terms of energy flows, and in those terms living systems do not appear to consistently increase their entropy, or at least their entropy production, through time. In fact, the contrary is often true (see chapters 2 and 3). Furthermore, to most biologists, it is not energy but structure that evolves, and thus it is not immediately apparent to them that there is a valid connection between thermodynamics and biological evolution.

In this regard, we had to grapple with the fundamental nature of entropy as a manifestation of dynamic processes. Specifically, is entropy primarily a manifestation of energy flows or of the passage of time? We noted Prigogine's (1980) comment that systems at equilibrium have no time parameter in their behavior. Thus, we decided to find the implications of considering entropy to be a general manifestation of the passage of time indicated to an observer by time-dependent, or irreversible, processes of all kinds. All time-dependent processes, under this view, should exhibit entropic behavior. The association of entropy strictly with energy flows would be an artifact of the types of systems used in studies first documenting the phenomenon. Thus, for example, if living systems do not show an increase in entropy due to energy flows, this does not mean that living systems are in any sense "negentropic." Rather, it means that energy is not the capacity that shows determinate behavior through time. In this way, we are able to see biological evolution and thermodynamic changes as special cases of a more general phenomenon of evolution, which we have been able to relate directly to cosmological considerations of existence in an expanding universe (Layzer 1975; Frautschi 1982). The second law is thus more than the natural law of energy flows; it is the natural law of history.

Evolutionary change in nonequilibrium systems can occur in two ways. In discussing the biological applications of each of these two models, we will concentrate on the way in which each explains the following points:

1. From where does the time asymmetry (cause of the macroscopic behavior) come? What is the *source* of this behavior and what is (are) the mechanism(s)?

2. How does biological information relate to mathematical considerations of macroscopic information—what we have called phenomenological information? How are organization and complexity distinguished? And, can biological information exhibit macroscopic behavior?

3. Can we use measures of statistical entropy (variations of Boltzmann's equation) to describe the behavior of biological systems?

The first class of models we call *boundary conditions models* (Bernstein et al. 1983; Hamilton 1977; Jantsch 1981; Johnson 1981; Wicken 1979). Prigogine and Stengers (1984) tried to restrict the name "thermodynamic" to refer only to models of this class. Such systems are inherently random and have no "memory" of initial conditions; macroscopic behavior is due to the imposition of forces from the boundaries (environment) of the system. If there are energy fluxes of sufficient magnitude flowing through the system, it can be "perturbed" from its inherently random state to a nonrandom state. Prigogine and Stengers referred to such production of nonrandom states as "self-organization," but we will call it *imposed organization*. Once enough entropy has been "exported" to the environment from the system, the boundary conditions force an irreversible evolution of the system, creating what Prigogine and Stengers term the *entropy barrier*. The evolution of nonrandom states from random initial conditions under this model, including production of macroscopic information in accordance with Layzer's formulation of the second law, is shown schematically in fig. 7.1.

The source of the time asymmetry in this model is energy fluxes from the environment of the system. Evolution is a historically contingent process in which evolving lineages "bounce around" a static

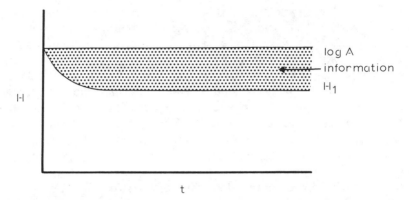

Fig. 7.1. Schematic representation of the evolution of nonrandom states in boundary conditions models of macroscopic behavior. (Redrawn from Brooks and Wiley 1985.)

phase space defining all possible configurations of the system according to the "selective" influence of the specific fluxes. A biological mechanism has not yet been described for this view. It is supposed to have the property of causing living systems to evolve in the direction of a selected number of ecological states. Biological concepts of "niche" are treated as direct analogs of quantum microstates. Thus, the mechanism is envisaged to have a combination of Darwinian and Lamarckian properties (see Ho and Saunders 1984).

Macroscopic information is defined as the difference between the maximum possible entropy and the entropy of the observed state of the system, in accordance with Layzer's formulation. It is a phenomenological description of the system at any point in time. However, since boundary conditions assume initial randomness and no inherent source of macroscopic behavior, there is no clear connection between phenomenological information and biological information. In fact, it is possible that the requirement of initial randomness implies that our perceptions of inherent biological information are artifactual, and all notions of genetic determinism should be abandoned. Furthermore, since information is only a phenomenological description of the system and not an inherent property of the system, it cannot logically exhibit macroscopic (entropic) behavior.

This class of models does not provide a clear distinction between organization and complexity. Wicken (1979) suggested that increasing organization be associated with higher entropy. However, Wicken used a measure of *microscopic information* (based on many-particle correlations derived from the theory of dilute gasses; for discussion see Layzer 1977) as a basis for his assertions. We can see from fig.

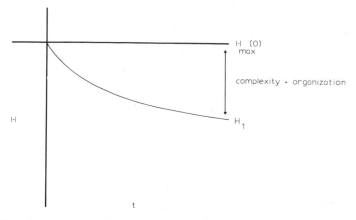

Fig. 7.2.   Organization and complexity are not clearly distinguished in macroscopic descriptions of boundary conditions models. (Redrawn from Brooks and Wiley 1985.)

7.1 that *macroscopic information,* the relevant parameter, increases only if entropy decreases. Thus, it has yet to be shown how organization and complexity can increase simultaneously without mixing microscopic and macroscopic considerations; conversely, it has not been shown how this mixture is justified. Therefore, we are left with the intuition that organization and complexity are contributing in some manner to the information component (fig. 7.2).

These models appear to be derived from considerations of quantum theory. If we assume the existence of a certain number of quantum microstates, called log A, the most probable distribution of the particles comprising the system is one in which all quantum states are filled randomly and equally. Any observed departure from randomness indicates the action of a selective force, in the form of energy fluxes, from the boundary conditions of the system. Since the quantum states are nonhistorical class constructs, particular configurations of such states are as well. The configuration of quantum states into which the system can be perturbed by the selective force will tend to be the most probable given the force(s). Since the system has no "memory" of initial conditions, the same selective force applied for a long enough period of time will perturb the system into the same configuration, no matter where it was initially. Thus, the *outcomes* of the interplay of matter and energy, that is, organisms and species, will be class constructs and not individualized systems. Under this view, macroscopic behavior is strictly a statistical artifact, and the use of statistical entropy measures is easily defended.

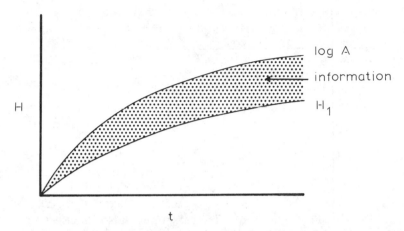

Fig. 7.3.   Schematic representation of the evolution of nonrandom states in initial conditions models of macroscopic behavior. (Redrawn from Brooks and Wiley 1985.)

These models have been applied to biology in order to provide a new way of explaining established biological concepts of fitness, selection, and adaptation. For this reason, we consider them to represent the search for a new mechanism to support an old theory. That mechanism has yet to be discovered. Van der Steen and Voorzanger (1984) concluded a discussion of the proposal by Bernstein et al. (1983) thus:

> The "basic dynamics" of Bernstein *et al.* is obviously a "general framework" merely in the sense of a heuristic device. It is a means of developing quite different laws of selection for different levels of organization. Accordingly, the "central" concept of selection will but pave the way for an array of very specific concepts as soon as "the" theory is really developed.

We call the second class of models the *initial conditions models*. This is a class of models Prigogine and Stengers (1984) call "dynamic" models. According to Prigogine and Stengers, such models are totally deterministic and not irreversible; that is, they are microscopic. However, Layzer (1975; see also Frautschi 1982) has shown that a subset of these models is macroscopic. If the maximum possible entropy of a system increases through time, and if any inherent constraints slow the rate at which the entropy of the system actually increases, macroscopic information will be produced (fig. 7.3). The observed entropy represents the degree to which the system has become random, given its initial conditions constraints. In this class of models, the system contains no microscopic information of significance to the macroscopic behavior of the system, but it exhibits macroscopic order initially.

Unlike the boundary conditions models, such systems "remember" at least some initial conditions, and thus information becomes a part of the causal makeup of the system. The macroscopic information at one point in time defines macroscopic order in the system that constrains, or informs, the future evolution of the system; note that this information does not *determine* the future. Information *emerges* from such behavior; it is not forced on the system. This we refer to as real *self-organization,* and we think it is in the spirit of such approaches as Haken's synergetics (Haken 1978; Shimizu and Haken 1983). Of course, the system must also function within an environmental (boundary condition) context, and this class of models does not rule out boundary condition effects. Nor does it rule out the possibility that environmental constraints can become initial condition constraints.

The source of time asymmetry in an initial conditions model applied to biology is the history of the system, at any functional level. Coherent biological functioning requires some degree of organization of matter. If all biological systems exhibit some degree of macroscopic order, and if that order is provided by the previous generation, there will be an inherent time asymmetry in the process. Biological inheritance ensures that the functioning products of reproduction are highly organized. This will be true even if the flow of inheritance is imperfect and even if boundary conditions affect the system as well. Thus, under this class of models, established mechanisms of inheritance are the primary cause of macroscopic behavior.

Phenomenological information for this class of models is defined in the same manner as for the boundary condition models. In this case, phenomenological information reflects the degree to which initial conditions constraints have kept actual entropy increases lower than maximal; it is an assessment of the degree to which the system has failed to become maximally random. That nonrandomness, or macroscopic order, which is transmitted to the next generation as initial conditions, *constrains, burdens* (Riedl 1978), *instructs* (Waddington 1977), or *informs* the system as to possible energy dissipation. In this sense, the information is more than a phenomenological description of the system—it is part of the causal makeup of the system. Information is thus a real capacity of the system in biological systems, something anticipated by Gatlin (1972). When cast in a

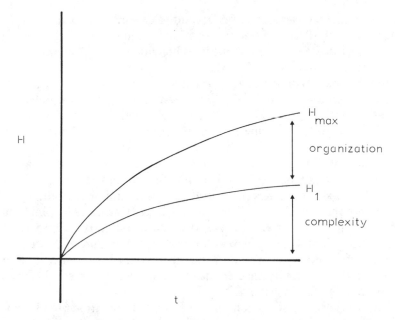

Fig. 7.4.   Organization and complexity are clearly distinguished in macroscopic descriptions of initial conditions models. (Redrawn from Brooks and Wiley 1985.)

genetic and epigenetic framework, the phenomenological information is a direct assessment of the effects of inherent biological information in any system having an inherited component. Inherent biological information is thus responsible for the phenomenological information. The amount of phenomenological information is at least partially a function of the system's entropy. Furthermore, that inherent biological information is responsible for the initial conditions constraints on biological macroscopic behavior, such as ontogeny or evolution. Thus, under this class of models, information exhibits macroscopic (entropic) behavior, a concept that is incoherent under boundary conditions models.

The initial conditions models also provide insight into the relationship between organization and complexity. Following Wicken's (1979) intuitions, we can equate organization with phenomenological information, and complexity with entropy (fig. 7.4). Unlike Wicken's treatment, however, these equivalences are achieved without introducing microscopic considerations; so this is a more consistent view. Furthermore, simultaneous increase in organization and complexity can be achieved without invoking extraordinary circumstances. Paradoxically, this support for Wicken's intuitions comes from initial conditions models and not from the boundary conditions models he espoused.

Initial conditions are generally perceived as either completely determining the behavior of the system or playing no role at all in macroscopic behavior. In the first case, the system is closed, unaffected by boundary conditions. In the second case, the system is totally open but the boundaries are, implicitly or explicitly, closed in terms of the phase space in which the system evolves. The initial conditions model presented herein assumes that both the system and its boundaries are open and can evolve. The inherent time asymmetry provided by inheritance of biological information guarantees that evolving lineages will be distributed over an increasing number of microstates through time. Furthermore, the phase space defining the maximum number of microstates the evolving lineage could occupy (maximum entropy) will also increase through time. Finally, as long as replication rates are higher than mutation rates, the phase space will always increase at a faster rate than the ensemble of microstates occupied by the evolving lineage. So long as there is partially filled phase space, any macroscopic behavior will be entropic (see chapter 2 and Elsasser 1983). That being the case, statistical estimates of entropy will be appropriate measures of the behavior of the system.

Let us now assume that an observer perceives the evolution of the system represented by the H line in fig. 7.4 from log A in the same figure. As evolution proceeds, the system recedes from the observer. Entropy appears to decrease unless the observer takes into account that his vantage point of log A is moving in the same direction as the system, but at a faster rate. This class of models thus appears to derive its ultimate justification from relativity theory. Our theory, based on this view, uses established mechanisms of inheritance and is "non-Darwinian" only to the extent of rejecting ecological determinism. It is not teleological or vitalistic as were theories of neo-Lamarckism or orthogenesis (see Bowler 1983), because the appearance of design comes from historical constraint, not the existence of a future goal. Thus, it appears to be a new theory for established mechanisms, in contrast with the boundary conditions models.

Considerations of nonequilibrium thermodynamics and biology have thus led biology to an interesting state of affairs. There is a "quantum" view and a "relativity" view, both of which claim to represent general causal explanations for biology. The quantum view is highly concordant with certain aspects of biological theory but has not yet found a mechanism. The relativity view is concordant with established mechanisms of inheritance, but it rejects the primacy of several major points of current theory (although it does not reject their existence). Why is it that we prefer the relativity view? The following is a list of items we think are explained better by the relativity view than by the quantum view:

1. It explains how biological systems can appear to be both deterministic (history; initial conditions constraints) and stochastic (inheritance is imperfect, and hence entropic; boundary conditions can affect evolution).

2. It explains how evolution can be entropic but appear to be negentropic.

3. It provides a common currency (macroscopic entropy and information) and dynamic (fig. 7.3) to link all functional levels of biology.

4. It identifies mechanisms causing macroscopic behavior in biology, and these mechanisms are empirically established.

5. It identifies the source of inherent biological information as the self-organizing capabilities of living things and provides a direct link with phenomenological information.

6. It provides a consistent view of organization and complexity, and of their mutual relationship, in biological systems.

7. It reaffirms the importance of history and of historical analysis in evolutionary biology, since historical constraints are synonymous with initial conditions constraints.

8. It is a theory of historical structuralism. It does not subordinate form to function in evolutionary explanations but predicts that structural changes will occur more often than functional changes in the course of evolution (see Ross 1972a, 1972b; Brooks, O'Grady, and Glen 1985). Consider fig. 7.5. The area beneath the H curve represents all actually realized information states during a given time sequence. The areas between the H and log A curves, the macroscopic information, can be seen retrospectively as representing the information states that were physically possible during a given time period, but were never passed on to subsequent generations, either because they were never produced or because they never reproduced. Anything above the log A curve is inaccessible to the evolving system at any time. For each time point, the explanation for the evolution of the present states must be drawn from other realized states, which are historical. Nonhistorical structuralism of the type

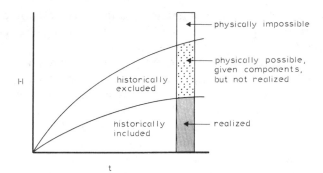

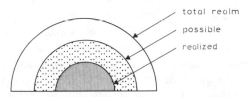

Fig. 7.5. History is an important component of the causal explanations in macroscopic systems constrained by initial conditions.

espoused by D'Arcy Thompson (1942) does not distinguish between historically excluded and physically excluded forms; any form that does not occur in nature is assumed to be physically excluded. This approach does not recognize the existence of an inherent log A curve. Historical functionalism, such as Darwinism and Lamarckism, on the other hand, does not distinguish between historically included and historically excluded forms. All unobserved possibilities are assumed to have evolved and then to have been selected against. This approach does not recognize the existence of an inherent H curve.

9. It offers a new view of the "balance of nature." It is a common perception that biological systems are organized by a balanced combination of opposing forces. For example, mutation is often considered a "randomizing" process opposed by selection; reproduction increases population size, which is opposed by competition. To some previous workers considering nonequilibrium thermodynamics and evolution, living systems are in a state of dynamic nonequilibrium precisely because these opposing forces operate simultaneously. Or-

ganization results from the dominance of the strongest force (e.g., the "slaving principle" of Haken 1978), so there is some essence of survival of the fittest under these views.

We have a different perspective, which can be couched as a question: How can it be that the determinate influences on evolution appear to be a set of opposing forces when in fact they are not opposing but complementary? The first clue to this question came from considering the evolutionary dynamic presented in chapter 2. Organization will emerge in any system characterized by coupled entropic phenomena proceeding at different rates. From the perspective of the faster phenomena, the slower phenomena appear to be moving in an opposite direction. Hence, relative to higher rates of mutation, the relatively low rates of changes in reproductive linkage patterns appear to be opposing the effects of mutation in sexually reproducing lineages. The end result is macroscopic organization of a type not found in asexual species. We have shown, however, that both changes in information and changes in cohesion are entropic phenomena.

10. The relativistic view does not appear to rule out any phenomena explained by the quantum view, although it may not accord them the same significance. Alternatively, the quantum view rules out initial conditions constraints (historical determinism) of any kind. Thus, the relativistic view is more general and seems to subsume the quantum view. What we have produced is a theory that provides a world view in which organization results from the historical interaction of complementary processes proceeding at different rates rather than one in which organization results from a battle of opposing forces. We have even tried to show that truly opposing forces are disorganizing (see chapter 6).

In physics, relativity theory subsumed Newtonian mechanics. In postulating that our theory represents a kind of relativity theory itself, we find ourselves in an analogous position. Darwin apparently relied heavily on Newtonian views of physics and of the philosophy of science in formulating his theory (e.g., see Ruse 1982). Does our theory stand with respect to Darwinism and neo-Darwinism as relativity theory stands to Newtonian mechanics? One theory may be replaced by another in two ways. It can be shown to be false, or it can be shown to be a special case of a new theory, which is preferred because it is more comprehensive. In other words, a theory may be eliminated or subsumed. Because biological evolution has occurred, and is occurring, and because selection and competition can affect biological systems, we do not consider the Darwinian and neo-Darwinian traditions false by any means. However, they are incomplete relative to our theory (just as ours may be incomplete relative to another theory), and their explanatory frameworks will not allow them to become more complete. Therefore, if our views represent a new theory, we must be able to show that the previous theories can be reduced to special cases of (subsumed by) our theory.

Churchland (1979) provided a useful formalization of the concept of intertheoretical reduction. He stated: "A successful reduction is a fellswoop proof of displaceability. You do not map the old onto the new, but show that the new contains an equipotent image of the old within it." We may use Churchland's formal notation to investigate the question of whether or not neo-Darwinisn can be shown to have been subsumed by our theory. Let T_o and T_n refer to the old and new theories, respectively. T_o is not deduced from T_n to show a reduction; rather, a subtheory, S_n, is deduced from T_n and then shown to be an equipotent image (*sensu* Churchland) of T_o within T_n. In this case, S_n represents biological functionality, selection, stochastic mutations and population biology, the reality of species and speciation, and ecological associations. This S_n is not a completely unaided consequence of T_n, because T_n (our theory) actually stresses other factors. However, S_n is a deduction of a modification of T_n, called T'_n, which is formulated to stress the connections between T_n and T_o. This is the main reason we have spent so much time investigating the aspects of evolutionary behavior that are equivalent under either theory. Since, as we have shown, neo-Darwinism is a diffuse belief system, any T_o we identify to use for reduction would better be termed T'_o. Thus, if a reduction has been accomplished, T_o is related to T_n through T'_o and T'_n, both of which contain S_n. This seems to be the case because S_n is identical in both theories but T'_n contains the added aspects of (1) constraints placed on variation by accessing history, (2) an explanation of the general dynamics of biological processes acting at several hierarchical levels, and (3) an account of, and explanation for, the differences between those aspects of biological processes which are reversible and those which are irreversible.

Neo-Darwinism is a relatively complete theory of proximal causes in evolution. Its three major principles, selection, competition, and dispersal, were formulated into a coherent theory by Darwin and have guided fruitful research for 125 years. But as a complete theory of *evolution*, neo-Darwinism asserts that evolution is strictly a matter of historical contingency. We can appreciate the sentiments that led some evolutionists to reject any form of ultimate cause in biology.

That is, if "ultimate cause" is equated with "final cause" or "teleological behavior," rather than with axiomatic causal behavior as well, formulation of a theory based only on proximal causes would exorcise the specter of teleology from biological explanations. But as O'Grady (1984) and Brooks and O'Grady (submitted for publication) have shown, proximal cause explanations are functionalist explanations, which are either incomplete as evolutionary explanations (ahistorical functionalism; descriptions of teleonomic behavior) or, paradoxically, teleological in logical form (historical functionalism; species evolve in order to perform the functions they do). Indeed, some recent evolutionists (Wicken 1980; Johnson 1981) have embraced teleology to a degree. Darwin referred to his theory as a theory of *natural selection*. In claiming that Darwinism is a complete theory of *evolution,* neo-Darwinists have unjustifiably extended proximal causes to the level of ultimate causes. This has produced an incomplete and relatively weak theory.

The essential incompleteness of neo-Darwinism as a theory of evolution lies in its passive exclusion of axiomatic causal behavior. We have tried to show that history exerts such an influence through the behavior of time-dependent processes, manifested by varying degrees of self-organization at different functional levels of biology. These are embodied in our four principles, irreversibility, individuality, inherent constraint, and compensatory changes. Our theory provides an axiom of ultimate causality, historically determined inherent directionality. The theory of natural selection nests within that more general theory of evolution as a theory of proximal causes affecting particular evolving lineages.

In the absence of proximal causes impinging on an evolving lineage, we would expect both complexity (entropy) and organization (information) to increase through time. The rate of entropy increase would be minimized and the degree of organization maximized through time; the lineage would become maximally self-organized. Proximal causes could affect the rate of entropy increase and the degree of organization. The biological system would change to an optimal accommodation to the extrinsic factors given the system's inherent constraints.

Churchland (1979) further asserted that cross-theoretical reduction does not require correspondence rules to preserve meaning in general and that cross-theoretical identity is not part of reduction proper. We have already seen that the analytical languages (systematic techniques) of our theory and of some neo-Darwinists (in the sense of T_o') have different intensional structures (chapter 5). That is, the semantically important systematic statements in each do not always correspond, and may diverge appreciably. And yet, all the syntactical elements are accounted for in both. The points at which correspondence does not preserve meaning are the points at which the two theories differ fundamentally. But this does not preclude one from being subsumed by the other.

Churchland stated further that the reduction of a T_o to a T_n would be accompanied by a reduction of T_o's ontology to T_n's ontology if, and only if, the reduction were sufficiently smooth to allow the relevant cross-theoretical identity claims to be sustained. Otherwise, the ontology of T_o will be a casualty of the reduction and will be replaced by the ontology of T_n. In this case, nongenealogical taxonomic grades have no ontological status under the new theory, nor do species as natural kinds into which organisms develop or evolve. The ontology of our theory is the ontology of individuals and historical groups.

We may also express a preference for the relativistic view over the quantum view on ontological grounds. Under a quantum view, macroscopic systems are class constructs comprising individual parts, none of which is a macroscopic system. Hence, individual cells are components of developing organisms, but the organism exhibits macroscopic behavior not directly attributable to its cells. Likewise, species are class constructs filled with individual organisms and ecosystems are class constructs filled with individual species.

The major ontological problems arising from such a view are problems of incompleteness. Individuality is assumed but not explained by this perspective. This could possibly be circumvented by asserting that individuality is a perceptual artifact (i.e., species really *are* classes), but that would mean that genealogical phenomena are also artifacts. The ontology of the quantum view excludes initial conditions effects and historical causality in order to explain macroscopic behavior as a boundary conditions phenomenon. Individuality is now "explained" but hard inheritance is not (because hard inheritance is an initial conditions phenomenon). The incompleteness is not solved, only transferred to a different level.

Allowing some levels of individuality in biology while maintaining the quantum view creates other problems of incompleteness as well as problems of inconsistency. Any allowed level of individuality would be, by definition, not a level of macroscopic behavior but a level of microscopic building blocks. For example, if species are individuals, then they cannot be dissipative structures (macroscopic systems) but only parts of macroscopic systems. Such a view is incomplete because it does not allow a smooth reduction through all functional levels—classes and individuals are interspersed. Furthermore, such a view is inconsistent since developing organisms are simultaneously macroscopic systems, individuals, and, by most biological accounts, the building blocks of species. How can organisms be individuals and be macroscopic systems, species be individuals and not be macroscopic systems, and ecosystems be classes and be macroscopic systems?

A more important aspect of the analogy with Newtonian mechanics and relativity theory is the question of what use our theory has. Relativity theory certainly added a dimension to physics that has been productive, and it rightfully can be asked what we think our effort will contribute to biology. We can answer this inquiry in both a weak sense and a strong sense.

First of all, we find ourselves, as practicing evolutionary biologists, in a time of widespread dissatisfaction with various parts of current evolutionary theory (for reviews see Ruse 1982; Brooks 1984). Thus, it can be argued that any serious attempt to redirect thinking along new lines should be viewed as useful because it may lead to substantive advances not directly anticipated by the formulators of the original argument.

In a strong sense, we think our theory provides concrete advances in a number of different aspects of evolutionary biology in addition to the ten points listed earlier. First, we have provided a theory of biological evolution that is explicitly canonical with cosmological evolution. We have also constructed an evolutionary theory that purports to explain evolutionary phenomena at all functional levels based on a single unified conceptual framework. This focuses the main thrust of evolutionary explanations on developmental biology and systematics, along with population biology, rather than almost exclusively on population biology as has been the case previously. This encourages a broader range of biologists, with a broader range of views, to participate actively in evolutionary biology.

By adopting an explicitly structuralist approach to evolution, we have discovered a number of surprising and, in some cases, counterintuitive things. We have been able to relate biological evolution to entropy without having to postulate any negentropic properties. This is, we discovered, because the currency of entropic behavior in living systems is information specifying structure, and not energy. We have characterized living systems at all functional levels in terms of four cornerstone principles. The principle of irreversibility asserts that biological evolution is one manifestation of the more general phenomenon of cosmological evolution. The principle of individuality amounts to an assertion that origins are essential to understanding evolution, and it also amounts to a rejection of the idea that species and monophyletic groups are natural kinds. The principle of intrinsic constraints asserts that evolutionary changes are nonrandom, orderly, and constrained by the past history of each species. And the principle of compensatory changes asserts that biological evolution has a conservative tendency that allows the same dynamics to produce both determinism, and new variations and options simultaneously. We have shown that biological information can be produced by entropic phenomena. This allows many aspects of evolutionary history and evolutionary processes to be tested directly and explicitly.

Our insights have not been gained without perturbing the established theoretical order. Population biology, the core of neo-Darwinism, embeds equally well within our theory. Previously, the success of population genetics was taken as de facto evidence that neo-Darwinism was the only plausible theory of evolutionary mechanism, and when one considers past alternatives, this conclusion seems entirely warranted. However, we have found that some things cited as causal agents of evolution are actually effects of other causal agents. This is true of ecological interactions in particular. Most, if not all, of the ecological factors cited by previous evolutionary biologists as driving forces of evolutionary change are retained in our theory as constraints on evolution or as disorganizing factors. Some things, such as developmental processes, which we were taught to think of as effects of natural selection, are actually causal agents of coherent evolutionary change that account for the nonrandom variations with which natural selection is presented.

We have discovered that some things which we thought required special biological explanations, such as developmental, phylogenetic, and ecological organization or an increase in variation through time, do not. Rather, they are the predicted class of outcomes of the kind of entropy systems that living things are. Finally, we have discovered that some things we thought required evolutionary explanations do not even exist as products of the evolutionary process. Among these are such functional groupings as reptiles, fishes, adaptive zones, and transspecific evolution.

The pragmatic aspect of whether or not our theory is considered a new view of biological evolution or just a reformulation of the consensus, depends on two things. First, to what extent is our presentation viewed as providing helpful explanations for findings that would seem to be anomalous under the current paradigm? For example, we provide a uniform mechanistic and evolutionary explanation for interchromosomal effects, for systematic congruence of larval and adult characters even in the absence of recapitulation, for the limits on selection, for the success of outgroup-based systematic techniques, and for the observation that ecological diversification lags far behind morphological diversification in evolution.

Second, virtually every new theoretical framework is constructed along lines dictated by attempts to explain particular phenomena. Thus, there will always be some data with which any new theory is canonical, because the theory was designed to explain those observations. This is not sufficient. Our theory must also make predictions, which can be of two general types. We must predict which areas of biological research will provide useful insights into evolutionary mechanisms. We have attempted to pinpoint those areas in the preceding chapters, focusing attention on developmental biology, including developmental genetics, on population biology, on systematics, and on historical ecology.

We must also make some predictions about the kinds of findings research guided by our theory will uncover. These may be general or specific predictions about occurrences or prohibitions (certain things that should never be found). Such deterministic predictions can be made for deterministic systems, but not for indeterminate systems. For example, the thermal conditions (ΔQ) characterizing any chemical reaction cannot be predicted from first principles. However, once a reaction has been run and its ΔQ measured, it is possible to make deterministic predictions about the ΔQ of any future reactions of the same type. Biological systems are a combination of deterministic and indeterminate factors; thus, it should be possible to make deterministic predictions about some aspects of evolution. If we are correct, however, the deterministic aspects of evolution are the historical components; so our explicit predictions are in fact retrodictions, or predictions about what we will find to have been the past configurations of various lineages. This may make our theory seem somewhat weak to those who are used to thinking of theories only in terms of deterministic behavior. But we submit that evolutionary theories in general will all suffer from the same "problem." We may also make general predictions stemming from the principle of compensatory changes, many of which we postulated in the preceding chapters. The extent to which the assertions we have made are considered novel predictions by other biologists may hinge on what each one requires from a theory.

Our view of biological evolution suggests that it is an inherently nonviolent, accommodating process. We believe that the Darwinian and neo-Darwinian views, with their emphasis on external forces, promote a violent conception of the world. Under our theory, a biologist can note that two organisms are different without having to ask which one is better (fitter). To better illustrate this point of difference between the two views, we have found it helpful to use some Aristotelian distinctions about processes (adapted from Jaki 1966).

Aristotle had an organic, or organismic, view of the cosmos, which certainly is in accord with our own. He divided the dynamic processes of the cosmos into *natural* and *violent* "kineses" or *efficient causes*. Natural kineses were inherent processes, derived from the nature of the substance and explainable in those terms. Natural kineses included (1) growth, (2) alteration (differentiation), (3) "place" (locomotion), and (4) genesis (coming into existence). Violent kineses, on the other hand, were processes or influences extrinsic to the substance being observed imposed on that substance and having no general explanation. *Natural efficient causes,* or kineses, differ from Aristotelian *natural necessity final causes* by virtue of having what we would call today a stochastic component. Natural necessity final causes, such as gravity, are completely deterministic. On the other hand, growth, differentiation, locomotion, and reproduction often proceed in ways that, despite their historical constraints, are at least partially unpredictable. We submit that our theory of biology is recognizable as a process of natural efficient cause. This may be the reason we think we can find general explanations for evolutionary processes, whereas previous discussions of evolutionary biology have asserted that no such generalizations exist (e.g., Popper 1965). In a general humanist sense, we believe it should be useful to seriously entertain the possibility that biological processes are inherently nonviolent. We have no strong metaphors to offer such as "survival of the fittest" or "Nature red in tooth and claw," and that is just fine with us. We would like nothing better than to make sure that violent human behavior could no longer be justified and condoned as an unalterable consequence of an evolutionary legacy.

· References ·

Abeloos, M. 1956. *Les Metamorphoses*. Paris: Collection Armand Collin.

Alberch, P. 1980. Ontogenesis and morphological diversification. *Amer. Zool.* 20:653–667.

———. 1982. Developmental constraints in evolutionary processes. In *Evolution and Development,* ed. J. T. Bonner, 313–332. New York: Springer-Verlag.

———. 1985. Problems with the interpretation of developmental sequences. *Syst. Zool.* 34:46–58.

Alberch, P., and J. Alberch. 1981. Heterochronic mechanisms of morphological diversification and evolutionary change in the Neotropical salamander, *Bolitoglossus occidentalis* (Amphibia: Plethodontidae). *J. Morphol.* 167:249–264.

Allen, P. M. 1981. The evolutionary paradigm of dissipative structures. In *The Evolutionary Vision. AAAS Selected Symposia,* ed. E. Jantsch, 25–72. Boulder, Colo.: Westview Press.

Baker, R. J., and J. W. Bickham. 1980. Karyotypic evolution in bats: Evidence of extensive and conservative chromosomal evolution in closely related taxa. *Syst. Zool.* 29:239–253.

Bard, J. B. L. 1981. A model for generating aspects of zebra and other mammalian coat patterns. *J. Theor. Biol.* 93:363–385.

Basharin, G. P. 1959. On a statistical estimate for the entropy of a sequence of independent variables. *Theory Probab. Appl.* 4:333–336.

Beatty, J., and W. L. Fink. 1979. [Review of] Simplicity. *Syst. Zool.* 28:643–651.

Bernstein, H., H. C. Byerly, F. A. Hopf, R. A. Michod, and G. K. Vemulapalli. 1983. The Darwinian dynamic. *Quart. Rev. Biol.* 58:185–207.

Blum, H. F. 1968. *Time's arrow and evolution*. 3rd ed. Princeton: Princeton Univ. Press.

Boltzmann, L. 1877. Uber die Beziehung eines allgemeine mechanischen Satzes zum zweiten Hauptsatzes der Warmetheorie. *Sitzungsber. Akad. Wiss. Wien, Math.-Naturwiss. Kl.* 75:67–73.

Boucot, A. J. 1975. Standing diversity of fossil groups in successive intervals of geologic times viewed in the light of changing levels of provincialism. *J. Paleontol.* 49:1105–1111.

————. 1978. Community evolution and rates of cladogenesis. In *Evolutionary Biology,* ed. M. K. Hecht, W. C. Steere, and B. Wallace, 545–655, vol. 11. New York: Plenum.

————. 1982. *Paleobiologic Evidence of Behavioral Evolution and Co-evolution.* Corvallis, Ore.: by the author.

————. 1983. Does evolution take place in an ecological vacuum? *J. Paleontol.* 37:1–30.

Bowler, P. J. 1983. *The Eclipse of Darwinism.* Baltimore: Johns Hopkins Univ. Press.

Bremer, K., and H.-E. Wanntorp. 1979. Hierarchy and reticulation in systematics. *Syst. Zool.* 28:624–626.

Bridgeman, P. W. 1945. Some general principles of operational analysis. *Psychol. Rev.* 52:246.

Brillouin, L. 1962. *Science and Information Theory.* New York: Academic Press.

Britten, R. J., and Davidson, E. H. 1969. Gene regulation for higher cells: A theory. *Science* 165:349–357.

Brooks, D. R. 1979. Testing the context and extent of host-parasite coevolution. *Syst. Zool.* 28:299–307.

————. 1980. Allopatric speciation and non-interactive parasite community structure. *Syst. Zool.* 29:192–203.

————. 1981a. Classifications as languages of empirical comparative biology. In *Advances in Cladistics: Proceedings of the First Meeting of the Willi Hennig Society,* ed. V. A. Funk and D. R. Brooks, 61–70. New York: New York Botanical Garden.

————. 1981b. Hennig's parasitological method: A proposed solution. *Syst. Zool.* 30:229–249.

————. 1982. A simulations approach to discerning possible sister-groups of *Dioecotaenia* Schmidt, 1969 (Cestoda: Tetraphyllidea: Dioecotaeniidae). *Proc. Helminthol. Soc. Wash.* 49:56–61.

————. 1984. What's going on in evolution? A brief guide to some new ideas in evolutionary theory. *Can. J. Zool.* 61:2637–2645.

————. 1985. Historical ecology: A new approach to studying the evolution of ecological associations. *Ann. Mo. Bot. Gard.*

Brooks, D. R., P. H. LeBlond, and D. D. Cumming. 1984. Information and entropy in a simple evolution model. *J. Theor. Biol.* 109:77–93.

Brooks, D. R., M. A. Mayes, and T. B. Thorson. 1981. Systematic review of cestodes infecting freshwater stingrays (Chondrichthyes: Potamotrygonidae) including four new species from Venezuela. *Proc. Helminthol. Soc. Wash.* 48:43–64.

Brooks, D. R., R. T. O'Grady, and D. R. Glen. 1985. Phylogenetic analysis of the Digenea (Platyhelminthes: Cercomeria) with comments on their adaptive radiation. *Can. J. Zool.* 63:411–443.

Brooks, D. R., and R. T. O'Grady. Submitted for publication. Nonequilibrium thermodynamics and different axioms of evolution.

Brooks, D. R., T. B. Thorson, and M. A. Mayes. 1981. Freshwater stingrays (Potamotrygonidae) and their helminth parasites: Testing hypotheses of evolution and coevolution. In *Advances in Cladistics: Proceedings of the First Meeting of the Willi Hennig Society,* ed. V. A. Funk and D. R. Brooks, 147–175. New York: New York Botanical Garden.

Brooks, D. R., and E. O. Wiley. 1984. Evolution as an entropic phenomenon. In *Evolutionary Theory: Paths to the Future,* ed. J. W. Pollard, 141–171. New York: Wiley.

———. 1985. Theories and methods in different approaches to phylogenetic systematics. *Cladistics* 1:1–11.

Brown, W. L., Jr., and E. O. Wilson. 1956. Character displacement. *Syst. Zool.* 5:49–64.

Brush, S. G. 1983. *Statistical Physics and the Atomic Theory of Matter, from Boyle and Newton to Landau and Onsager.* Princeton: Princeton Univ. Press.

Bush, G. L. 1975a. Sympatric speciation in phytophagus parasitic insects. In *Evolutionary Strategies of Parasitic Insects and Mites,* ed. P. W. Price, 187–206. New York: Plenum Press.

———. 1975b. Modes of animal speciation. *Ann. Rev. Ecol. Syst.* 6:339–364.

Bush, G. L., S. M. Case, A. C. Wilson, and J. L. Patton. 1977. Rapid speciation and chromosomal evolution in mammals. *Proc. Nat. Acad. Sci.* 74:3942–3946.

Chaitin, G. J. 1975. Randomness and mathematical proof. *Sci. Amer.* 232:47–52.

Chambers, R. 1844. *Vestiges of the Natural History of Creation.* London: Leicester Univ. Press.

Charlesworth, B., R. Lande, and M. Slatkin. 1982. A neo-Darwinian commentary on macroevolution. *Evolution* 36:474–498.

Chomsky, N. 1965. *Aspects of the Theory of Syntax.* Cambridge, Mass.: M.I.T. Press.

Churchland, P. M. 1979. *Scientific Realism and the Plasticity of Mind.* Cambridge: Cambridge Univ. Press.

———. 1982. Is *Thinker* a natural kind? *Dialogue* 21:223–238.

Collier, J. in press. Entropy in evolution. *Biol. and Philos.* 1.

Conklin, E. G. 1896. Discussion of the factors of organic evolution from the embryological standpoint. *Proc. Amer. Philos. Soc.* 35:78–88.

Crisci, J. V., and T. F. Steussy. 1980. Determining primitive character states for phylogenetic reconstruction. *Syst. Bot.* 5:112–135.

Croizat, L. 1962. *Space, Time, Form. The Biological Synthesis.* Caracas, Venezuela: by the author.

Croizat, L., G. Nelson, and D. E. Rosen. 1974. Centers of origin and related concepts. *Syst. Zool.* 23:265–287.

Crow, J. F., and M. Kimura. 1970. *An Introduction to Population Genetics Theory.* New York: Harper and Row.

D'Arcy Thompson, W. 1942. *On Growth and Form*. Cambridge: Cambridge Univ. Press.

Darwin, C. 1859. *On the Origin of Species*. London: J. Murray.

Deardorff, T. L., D. R. Brooks, and T. B. Thorson. 1981. Two species of *Echinocephalus* (Nematoda: Gnathostomidae) from Neotropical stingrays. *J. Parasitol.* 67:433–439.

de Beer, G. 1958. *Embryos and Ancestors*. Oxford: Oxford Univ. Press.

de Moivre, A. 1756. *The Doctrine of Chances: Or, a Method of Calculating the Probability of Events in Play*. 2nd ed. Photo reproduction, 1967. New York: Chelsea Press.

Denbigh, K. G. 1975. A non-conserved function for organized systems. In *Entropy and Information in Science and Philosophy,* ed. L. Kubat and J. Zemen, 83–92. New York: American Elsevier.

Dobzhansky, T. 1937. *Genetics and the Origin of Species*. New York: Columbia Univ. Press.

———. 1951. *Genetics and the Origin of Species*. 3rd ed. New York: Columbia Univ. Press.

———. 1970. *Genetics of the Evolutionary Process*. New York: Columbia Univ. Press.

Dobzhansky, T., F. J. Ayala, G. L. Stebbins, and J. W. Valentine. 1977. *Evolution*. San Francisco: W. H. Freeman.

Domning, D. P. 1981. Sea cows and sea grasses. *Paleobiology* 7:417–420.

Dover, G. A. 1982. Molecular drive: A cohesive mode of species evolution. *Nature* 229:111–117.

Edelman, G. M. 1984a. Cell-adhesion molecules: A molecular basis for animal form. *Sci. Amer.* 250:118–129.

———. 1984b. Cell adhesion and morphogenesis: The regulator hypothesis. *Proc. U.S. Nat. Acad. Sci.* 81:1460–1464.

Ehrlich, P., and P. H. Raven. 1969. Differentiation of populations. *Science* 165:1228–1232.

Eigen, M. 1971. Self-organization of matter and the evolution of biological macromolecules. *Naturwissenschaften* 58:465–522.

Eldredge, N., and J. Cracraft. 1980. *Phylogenetic Patterns and the Evolutionary Process*. New York: Columbia Univ. Press.

Eldredge, N., and S. J. Gould. 1972. Punctuated equilibria: An alternative to phyletic gradualism. In *Models in Paleobiology,* ed. T. J. M. Schopf, 130–145. San Francisco: Freeman, Cooper.

Elias, P. 1983. Entropy and the measure of information. In *The Study of Information,* ed. F. Machlup and U. Mansfield, 497–502. New York: Wiley-Interscience.

Elsasser, W. M. 1983. Biological application of the statistical concepts used in the Second Law. *J. Theor. Biol.* 105:103–116.

Endler, J. A. 1977. *Geographic Variation, Speciation, and Clines*. Princeton: Princeton Univ. Press.

Engels, W. R. 1983. The P family of transposable elements in *Drosophila*. *Ann. Rev. Genet.* 17:315–344.

Erneux, T., and J. Hiernaux. 1979. Chemical patterns in circular morphogenetic fields. *Bull. Math. Biol.* 41:461–468.

Estabrook, G. F. 1972. Cladistic methodology: A discussion of the theoretical basis for the induction of evolutionary history. *Ann. Rev. Ecol. Syst.* 3:427–456.

———. 1978. Some concepts for the estimation of evolutionary relationships in systematic botany. *Syst. Bot.* 3:146–158.

Evans, R. B. 1966. Basic relationships among entropy, exergy, energy and availability. In *Principles of Desalination,* ed. K. S. Spiegler, 44–66. New York: Academic Press.

Faeth, S. H. 1984. Density compensation in vertebrates and invertebrates: A review and an experiment. In *Ecological Communities: Conceptual Issues and the Evidence,* ed. D. R. Strong, Jr., D. Simberloff, L. G. Abele, and A. B. Thistle, 491–509. Princeton: Princeton Univ. Press.

Farris, J. S. 1970. Methods for computing Wagner trees. *Syst. Zool.* 19:83–92.

———. 1979. The information content of the phylogenetic system. *Syst. Zool.* 28:483–519.

Farris, J. S., A. G. Kluge, and M. J. Eckardt. 1970a. A numerical approach to phylogenetic systematics. *Syst. Zool.* 19:172–189.

———. 1970b. On predictivity and efficiency. *Syst. Zool.* 19:363–372.

Fink, W. L. 1982. The conceptual relationship between ontogeny and phylogeny. *Paleobiology* 8:254–264.

Frautschi, S. 1982. Entropy in an expanding universe. *Science* 217:593–599.

Funk, V. A. 1981. Special concerns in estimating plant phylogenies. In *Advances in Cladistics: Proceedings of the First Meeting of the Willi Hennig Society,* ed. V. A. Funk and D. R. Brooks, 73–86. New York: New York Botanical Garden.

Futuyma, D. J. 1979. *Evolutionary Biology.* Sunderland, Mass.: Sinauer Assoc.

Galau, G. A., W. H. Klein, M. M. Davis, B. J. Wold, R. J. Britten, and E. H. Davidson. 1976. Structural gene sets active in embryos and adult tissues of the sea urchin. *Cell* 7:487–505.

Garstang, W. 1922. The theory of recapitulation: A critical restatement of the biogenetic law. *Zool. J. Linnean Soc. Lond.* 35:81–101.

Gatlin, L. L. 1972. *Information Theory and the Living System.* New York: Columbia Univ. Press.

Ghiselin, M. T. 1966. On psychologism in the logic of taxonomic controversies. *Syst. Zool.* 15:207–215.

———. 1974. A radical solution to the species problem. *Syst. Zool.* 23:536–544.

———. 1980. The failure of morphology to assimilate Darwinism. In *The Evolutionary Synthesis:* Perspectives on the Unification of Biology, ed. E. Mayr and W. B. Provine, 180–193. Cambridge, Mass.: Harvard Univ. Press.

————. 1981. Categories, life and thinking. *Behav. Brain Sci.* 4:269–313 (with peer comments).

Gibbs, J. W. 1902. *Elementary Principles in Statistical Mechanics: Developed with Especial Reference to the Rational Foundation of Thermodynamics.* New York: Scribner's.

Gilbert, J. J. 1980. Developmental polymorphism in the rotifer *Asplanchna sieboldi. Amer. Sci.* 68:636–646.

Gilmartin, A. J. 1983. Evolution of mesic and xeric habits in *Tillandsia* and *Vriesia* (Bromeliaceae). *Syst. Bot.* 8:233–242.

Ginzburg, L. R. 1980. Ecological implications of natural selection. In *Lecture Notes in Biomathematics 39. Vito Volterra Symposium on Mathematical Models in Biology,* ed. C. Barigozzi, 171–183. Berlin: Springer-Verlag.

Goldschmidt, R. B. 1940. *The Material Basis of Evolution.* New Haven: Yale Univ. Press.

————. 1952. Evolution as viewed by one geneticist. *Amer. Sci.* 40: 84–98.

Goodwin, B. C. 1982. Development and evolution. *J. Theor. Biol.* 97:43–55.

Goodwin, B. C., and L. E. H. Trainor. 1983. The ontogeny and phylogeny of the pentadactyl limb. In *Development and Evolution,* ed. B. C. Goodwin, N. Holder, and C. G. Wylie, 75–98. Cambridge: Cambridge Univ. Press.

Gould, S. J. 1980. Is a new and general theory of evolution emerging? *Paleobiology* 6:119–130.

Gould, S. J., and E. S. Vrba. 1982. Exaptation–a missing term in the science of form. *Paleobiology* 8:4–15.

Grant, V. 1971. *Plant Speciation.* New York: Columbia Univ. Press.

Greene, M. 1978. Individuals and their kind: Aristotelian foundations of biology. In S. F. Spicker, ed., *Organism, Medicine, and Metaphysics,* pp. 121–36. Dordrecht, Netherlands: Reidel.

Griffiths, G. C. D. 1974. On the foundations of biological systematics. *Acta Biotheor.* 23:85–131.

Haeckel, E. 1866. *Generelle Morphologie der Organismen.* Berlin: G. Reimer.

————. 1877. Studien zur Gasträa - Theorie. *Biol. Stud.* 2:171–226.

Haken, H. 1978. *Synergetics.* Berlin: Springer-Verlag.

Hamburger, V. 1980. Embryology and the modern synthesis in evolutionary theory. In *The evolutionary synthesis: Perspectives on the Unification of Biology,* ed. E. Mayr and W. B. Provine, 96–112. Cambridge, Mass.: Harvard Univ. Press.

Hamilton, H. J. 1977. A thermodynamic theory of the origin and hierarchical evolution of living systems. *Zygon* 12:289–335.

Harlan, J. R., and J. M. J. deWet. 1963. The compilospecies concept. *Evolution* 17:497–501.

Harre, R. 1972. *The Philosophies of Science.* Oxford: Oxford Univ. Press.

Harris, W. F., and R. O. Erikson. 1980. Tubular arrays of spheres: Geometry, continuous and discontinuous contraction, and the role of moving dislocations. *J. Theor. Biol.* 83:215–246.

Harrison, L. G. 1981. Physical chemistry of biological morphogenesis. *Chem. Soc. Rev.* 10:491–528.

———. 1982. An overview of kinetic theory in developmental modeling. In *Developmental Order: Its Origin and Regulation,* ed. S. Subtelny, 3–33. New York: Alan R. Liss.

Harrison, L. G., and T. C. Lacalli. 1978. Hyperchirality: A mathematically convenient and biochemically possible model for the kinetics of morphogenesis. *Proc. Roy. Soc. Lond. B* 202:361–397.

Harvey, P. H., and M. Slatkin. 1982. Some like it hot: Temperature-determined sex. *Nature* 296:807–808.

Hempel, C. G. 1966. *Philosophy of Natural Science.* Englewood Cliffs, N.J.: Prentice-Hall.

Hennig, W. 1950a. *Grundzuge einer theorie der phylogenetischen systematik.* Berlin: Deutscher Zentralverlag.

———. 1950b. *Die Larvenformen der Dipteren. Eine Uberricht uber die bisher bekannten Jugendstudien der Zweiflugeligen Insekten.* Berlin: Deutsches Verlag.

———. 1953. Kritische Bemerkungen zum phylogenetischen System der Insekten. *Beitr. Entomol.* 3:1–85.

———. 1966. *Phylogenetic Systematics.* Urbana: Univ. of Illinois Press.

———. 1969. *Die Stammesgeschichte der Insekten.* Frankfurt: W. Kramer.

Hertwig, O. 1906. Uber die Stellung der vergleichenden Entwicklungslehre zur vergleichenden Anatomie, zur Systematik und Deszendenztheorie. *Handbuch Vergl. Exper. Entwicklungslehre Wirbeltieres III* 3:149–183.

Hill, J. E., and J. D. Smith. 1984. *Bats: A Natural History.* London: British Museum (Natural History).

His, W. 1874. *Unsere Korperform und das physiologische Problem inherer Entstehung.* Leipzig: Vogel.

Ho, M.-W., and P. T. Saunders. 1979. Beyond neo-Darwinism—an epigenetic approach to evolution. *J. Theor. Biol.* 78:573–591.

Ho, M.-W., and P. T. Saunders, ed. 1984. *Beyond Neo- Darwinism: An Introduction to the New Evolutionary Paradigm.* London: Academic Press.

Hollinger, H. B., and M. J. Zenzen. 1982. An interpretation of macroscopic irreversibility within the Newtonian framework. *Philos. Sci.* 49:309–354.

Holmes, J. C. 1971. Two new sanguinicolid blood flukes (Digenea) from scorpaenid rockfishes (Perciformes) of the Pacific coast of North America. *J. Parasitol.* 57:209–216.

———. 1973. Site selection by parasitic helminths: Interspecific interactions, site segregation, and their importance to the development of helminth communities. *Can. J. Zool.* 51:333–347.

Holmes, J. C., and P. W. Price. 1980. Parasite communities: The roles of phylogeny and ecology. *Syst. Zool.* 29:203–213.

Holmes, S. J. 1944. Recapitulation and its supposed causes. *Quart. Rev. Biol.* 19:319–331.

Horder, J. S. 1981. On not throwing the baby out with the bath water. In *Evolution Today,* ed. G. G. E. Scudder and J. L. Reveal, 163–180. Pittsburgh: Hunt Institute for Botanical Documentation, Carnegie-Mellon University.

Howden, H. F. 1982. Larval and adult characters of *Frickius* Germain: Its relationship to the Geotrupini, and a phylogeny of some major taxa in the Scarabaeoidea (Insecta: Coleoptera). *Can. J. Zool.* 60:2713–2724.

Hull, D. L. 1976. Are species really individuals? *Syst. Zool.* 25:174–191.

———. 1980. Individuality and selection. *Ann. Rev. Ecol. Syst.* 11:311–332.

———. 1981. Metaphysics and common usage. *Behav. Brain Sci.* 4:290–291.

———. 1983. Darwin and the nature of science. In *Evolution from Molecules to Man,* ed. D. S. Bendall, S. C. Metcalfe, and N. J. Jardine, 63–80. Cambridge: Cambridge Univ. Press.

Humphries, C. J. 1983. Primary data in hybrid analysis. In *Advances in Cladistics,* ed. N. I. Platnick and C. J. Humphries, 89–104, vol. 2. New York: Columbia Univ. Press.

Hutcheson, K. 1970. A test for comparing diversities based on the Shannon formula. *J. Theor. Biol.* 29:151–154.

Huxley, J. 1944. *Evolution: The Modern Synthesis.* London: George Allen.

Jablonski, D., J. J. Sepkoski, Jr., D. J. Bottjer, and P. M. Sheehan. 1983. Onshore-offshore patterns in the evolution of Phanerozoic shelf communities. *Science* 222:1123–1125.

Jacob, F., and J. Monod. 1961. Genetic regulatory mechanism in the synthesis of proteins. *J. Molec. Biol.* 2:318–356.

Jaki, S. L. 1966. *The Relevance of Physics.* Chicago: Univ. of Chicago Press.

Jantsch, E., ed. 1981. *The Evolutionary Vision. AAAS Selected Symposia.* Boulder, Colo.: Westview Press.

Jaynes, E. T. 1957a. Information theory and statistical mechanics. *Phys. Rev.* 106(2nd series):620–630.

———. 1957b. Information theory and statistical mechanics. *Phys. Rev.* 108(2nd series):171–190.

Johnson, L. 1981. The thermodynamic origin of ecosystems. *Can. J. Fish. Aquat. Sci.* 38:571–580.

Karreman, G. 1955. Topological information content and chemical reactions. *Bull. Math. Biophys.* 17:279–285.

Katz, M. J., and W. Goffman. 1981. Preformation of ontogenetic patterns. *Philos. Sci.* 48:438–490.

Kauffman, S. A. 1969. Metabolic stability and epigenesis in randomly constructed genetic nets. *J. Theor. Biol.* 22:437–467.

———. 1971. Gene regulation networks: A theory for their global structure and behavior. *Curr. Top. Devel. Biol.* 6:145–182.

————. 1973. Control circuits for determination and transdetermination. *Science* 181:310–318.

————. 1974. The large scale structure and dynamics of gene control circuits: An ensemble approach. *J. Theor. Biol.* 44:167–190.

————. 1977. Chemical patterns, compartments, and a binary epigenetic code in *Drosophila. Amer. Zool.* 17:631–648.

————. 1981. Pattern formation in the *Drosophila* embryo. *Philos. Trans. Roy. Soc. Lond. B* 295:567–594.

————. 1983. Developmental constraints: Internal factors in evolution. In *Development and Evolution,* ed. B. C. Goodwin, N. Holder, and C. G. Wylie, 195–225. Cambridge: Cambridge Univ. Press.

Kauffman, S.A., R. M. Shymko, and K. Trabert. 1978. Control of sequential compartment formation in *Drosophila. Science* 199:259–270.

Kettlewell, H. B. D. 1955. Selection experiments on industrial melanism in the Lepidoptera. *Heredity* 9:323–342.

————. 1961. The phenomenon of industrial melanism in the Lepidoptera. *Ann. Rev. Entomol.* 6:245–262.

————. 1965. Insect survival and selection for pattern. *Science* 148:1290–1296.

Kidwell, M. G. 1983. Intraspecific hybrid sterility. In *The Genetics and Biology of Drosophila,* ed. M. Ashburner, H. C. Carson, and J. N. Thompson, Jr., 125–154, vol. 3c. New York: Academic Press.

Kimura, M., and T. Ohta. 1971. *Theoretical Aspects of Population Genetics.* Princeton: Princeton Univ. Press.

King, A. W., and S. L. Pimm. 1983. Complexity, diversity and stability: A reconciliation of theoretical and empirical results. *Amer. Natur.* 122:229–239.

Kluge, A. G. 1985. Ontogeny and phylogenetic systematics. *Cladistics* 1:13–27.

Kluge, A. G., and J. S. Farris. 1969. Quantitative phyletics and the evolution of anurans. *Syst. Zool.* 18:1–32.

Kluge, A. G., and W. C. Kerfoot. 1973. The predictability and regularity of character divergence. *Amer. Natur.* 107:426–442.

Koch, A. L. 1984. Evolution vs. the number of gene copies per primitive cell. *J. Molec. Evol.* 20:71–76.

Kollar, E. J., and C. Fisher. 1980. Tooth induction in chick epithelium: Expression of quiescent genes for enamel synthesis. *Science* 207:993–995.

Kopaska-Merkel, D. C., and R. Haack. 1982. Herbert Spencer's theorem: A reply. *Syst. Zool.* 31:95–97.

Kucias, I. 1984. Uber den fur die Zellbiologie heuristischen Wert des relativen Strukturinformationgehalts, der Negentropie unter der negativen Entropie der Zelle. III. Der Zusammenhang zwischen der strukturellen Ordnung und der Entropie der Eukarotenzelle. *Biol. Zbl.* 103:123–138.

Lacalli, T. C., and L. G. Harrison. 1979. Turing's conditions and the analysis of morphogenetic models. *J. Theor. Biol.* 76:419–436.

Lamarck, J. B. 1803. Zoological philosophy. Translated by H. Elliot. Reprint 1963. New York: Hafner.

Landsberg, P. T. 1984a. Is equilibrium always an entropy maximum? *J. Stat. Physics* 35:159–169.

Landsberg, P. T. 1984b. Can entropy and "order" increase together? *Physics Letters* 102A:171–173.

Landsberg, P. T., and D. Tranah. 1980. Entropies need not be concave. *Physics Letters* 78A:219–220.

Lauder, G. V. 1981. Form and function: Structural analysis in evolutionary biology. *Paleobiology* 7:430–442.

———. 1982. Historical biology and the problem of design. *J. Theor. Biol.* 97:57–68.

Layzer, D. 1975. The arrow of time. *Sci. Amer.* 233:56–69.

———. 1977. Information in cosmology, physics and biology. *Int. J. Quantum Chem.* 12(suppl. 1):185–195.

———. 1978. A macroscopic approach to population genetics. *J. Theor. Biol.* 73:769–788.

———. 1980. Genetic variation and progressive evolution. *Amer. Natur.* 115:809–826.

Levin, L. A. 1984. Multiple patterns of development in *Streblospio benedicti* Webster (Spionidae) from three coasts of North America. *Biol. Bull.* 166:494–508.

Lewin, R. 1982. Biology is not postage stamp collecting. *Science* 216:718–720.

———. 1983. Santa Rosalia was a goat. *Science* 221:636–639.

Lewontin, R. C. 1974. *The Genetic Basis of Evolutionary Change.* New York: Columbia Univ. Press.

Lillie, F. R. 1898. Adaptation and cleavage. *Biol. Lect. MBL (Woods Hole)* 1898:43–67.

Lloyd, M., J. H. Zar, and J. R. Karr. 1968. On the calculation of information-theoretical measures of diversity. *Amer. Midl. Nat.* 79:257–272.

Lotka, A. J. 1924. *Elements of Mathematical Biology.* Reprinted 1956. New York: Dover Books.

Løvtrup, S. 1974. *Epigenetics: A treatise on theoretical biology.* New York: Wiley-Interscience.

———. 1981. The epigenetic utilization of the genomic message. In *Evolution Today,* ed. G. G. E. Scudder and J. L. Reveal, 145–161. Pittsburgh: Hunt Institute for Botanical Documentation, Carnegie-Mellon University.

Lucchesi, J. C. 1976. Interchromosomal effects. In *The Genetics and Biology of Drosophila melanogaster,* ed. M. Ashburner and E. Novitski, 315–329, vol. 1a. New York: Academic Press.

———. 1977. Dosage compensation: Transcription-level regulation of x-linked genes in *Drosophila. Amer. Zool.* 17:685–693.

Lucchesi, J. C., and D. T. Suzuki. 1968. The interchromosomal control of recombination. *Ann. Rev. Genet.* 2:53–86.

Lundberg, J. G. 1972. Wagner networks and ancestors. *Syst. Zool.* 21:398–413.

Lurie, D., and J. Wagensberg. 1979. Non-equilibrium thermodynamics and biological growth and development. *J. Theor. Biol.* 78:241–250.

MacArthur, R. H., and E. O. Wilson. 1967. *The Theory of Island Biogeography.* Princeton: Princeton Univ. Press.

McCoy, E. D., and K. L. Heck, Jr. 1976. Biogeography of corals, seagrasses, and mangroves: An alternate to the center of origin concept. *Syst. Zool.* 25:201–210.

———. 1983. Centers of origin revisited. *Paleobiology* 9:17–19.

Maddison, W. P., M. J. Donoghue, and D. R. Maddison. 1984. Outgroup analysis and parsimony. *Syst. Zool.* 33:83–103.

Maienschein, J. 1978. Cell lineage, ancestral reminiscence, and the biogenetic law. *J. Hist. Biol.* 11:129–158.

Manter, H. W. 1954. Some digenetic trematodes from fishes of New Zealand. *Trans. Roy. Soc. N.Z.* 82:475–568.

May, R. M. 1983. [Review of] Food webs. *Science* 220:295–296.

Maynard Smith, J. M. 1966. Sympatric speciation. *Amer. Natur.* 100:637–650.

———. 1970. Time in the evolutionary process. *Studium Generale* 23:266–272.

———. 1978. *The Evolution of Sex.* Cambridge: Cambridge Univ. Press.

Mayr, E. 1942. *Systematics and the Origin of Species.* New York: Columbia Univ. Press.

———. 1963. *Animal Species and Evolution.* Cambridge, Mass.: Harvard Univ. Press.

———. 1969. *Principles of Systematic Zoology.* New York: McGraw-Hill.

———. 1974. Teleological and teleonomic; A new analysis. In *Methodological and Historical Essays in the Natural and Social Sciences,* ed. R. S. Cohen and M. W. Wartofsky. Vol. 14, 91–117. *Boston Studies in the Philosophy of Science.* Boston: D. Reidel.

———. 1976. *Evolution and the Diversity of Life.* Cambridge, Mass.: Harvard Univ. Press.

———. 1978. Evolution. *Sci. Amer.* 239:46–55.

———. 1981. Biological classification: Toward a synthesis of opposing methodologies. *Science* 214:510–516.

———. 1982. How to carry out the adaptationist program? *Amer. Natur.* 121:324–334.

Mercer, E. H. 1981. *The Foundations of Biological Theory.* New York: Wilcy-Interscience.

Mitcheson, G. J. 1977. Phyllotaxis and the Fibonacci series. *Science* 196:270–275.

Montrol, E. W. 1983. The entropy function in complex systems. In *The Study of Information,* ed. F. Machlup and U. Mansfield, 503–511. New York: Wiley- Interscience.

Morowitz, H. J. 1968. *Energy Flow in Biology: Biological Organization as a Problem in Thermal Physics*. New York: Academic Press.

Murray, J. D. 1981. On pattern formation mechanisms for lepidopteran wing patterns and mammalian coat markings. *Philos. Trans. Roy. Soc. Lond.* B 295:473–496.

Naef, A. 1917. Die individuelle Entwicklung organischen Formen als Urkunde ihrer Stammesgeschichte. Jena.

———. 1931. Phylogenie der Tiere. Handbuch Vererbungswiss. 3:1–200.

Nei, M. 1965. Variation and covariation of gene frequencies in subdivided populations. *Evolution* 19:256–258.

Nelson, G. 1983. Reticulation in cladograms. In *Advances in Cladistics*, ed. N. I. Platnick and C. J. Humphries, 105–111. New York: Columbia Univ. Press.

Nelson, G., and N. I. Platnick. 1981. *Systematics and Biogeography: Cladistics and vicariance*. New York: Columbia Univ. Press.

O'Grady, R. T. 1982. Nonequilibrium evolution and ontogeny. *Syst. Zool.* 31:503–511.

———. 1984. Evolutionary theory and teleology. *J. Theor. Biol.* 107:563–578.

———. 1985. The phylogenetics of parasitic flatworm life cycles. *Cladistics* 1:159–170.

Ohta, T., and G. A. Dover. 1984. The cohesive population genetics of molecular drive. *Genetics* 108:501–521.

Oster, G., and P. Alberch. 1982. Evolution and bifurcation of developmental programs. *Evolution* 36:444–459.

Papentin, F. 1980. On order and complexity. I. General considerations. *J. Theor. Biol.* 87:421–456.

Paterson, H. E. H. 1978. More evidence against speciation by reinforcement. *South African J. Sci.* 74:369–371.

Patterson, C. 1978. Verifiability in systematics. *Syst. Zool.* 27:218–221.

Pechenik, J. A., and G. M. Lima. 1984. Relationship between growth, differentiation, and length of larval life for individually reared larvae of the marine gastropod, *Crepidula fornicata*. *Biol. Bull.* 166:537–549.

Pielou, E. C. 1974. *Population and Community Ecology: Principles and Methods*. New York: Gordon and Breach.

Pimm, S. L. 1982. *Food Webs*. London: Chapman and Hall.

———. 1984. The complexity and stability of ecosystems. *Nature* 307:321–326.

Pinto, J. D. 1984. Cladistic and phenetic estimates of relationship among genera of eupomphine beetles (Coleoptera: Meloidae). *Syst. Entomol.* 9:165–182.

Platnick, N. I., and H. D. Cameron. 1977. Cladistic methods in textual, linguistic and phylogenetic analysis. *Syst. Zool.* 26:380–385.

Platnick, N. I., and G. Nelson. 1978. A method of analysis for historical biogeography. *Syst. Zool.* 27:1–16.

Polikoff, D. 1981. C. H. Waddington and modern evolutionary theory. *Evol. Theory* 5:143–168.

Popper, K. R. 1965. *The Poverty of Historicism.* New York: Harper and Row.

Price, P. W. 1977. General concepts of the evolutionary biology of parasites. *Evolution* 3:405–420.

———. 1980. *Evolutionary Biology of Parasites.* Princeton: Princeton Univ. Press.

Prigogine, I. 1947. *Etude thermodynamique des phenomenes irreversibles.* Liege: Desoer.

———. 1967. Dissipative structures in chemical systems. In *Fast Reactions and Primary Processes in Chemical Kinetics,* ed. S. Claesson, 371–382. New York: Interscience.

———. 1980. *From Being to Becoming.* San Francisco: W. H. Freeman.

Prigogine, I., G. Nicolis, and A. Babloyantz. 1972. Thermodynamics of evolution. *Physics Today* 25(11):23–28, 25(12):38–44.

Prigogine, I., and I. Stengers. 1984. *Order Out of Chaos.* New York: Bantam.

Prigogine, I., and J. M. Wiame. 1946. Biologie et thermodynamique des phenomenes irreversibles. *Experientia* 2:451–453.

Rachootin, S. P., and K. S. Thompson. 1981. Epigenetics, paleontology, and evolution. In *Evolution Today,* ed. G. G. E. Scudder and J. L. Reveal, 181–193. Pittsburgh: Hunt Institute for Botanical Documentation, Carnegie-Mellon University.

Raff, R. A., and T. C. Kaufman. 1983. *Embryos, Genes and Ancestors.* New York: MacMillan.

Raup, D. M. 1968. Theoretical morphology of echinoid growth. *J. Paleontol.* 42:50–63.

Raup, D. M., and J. J. Sepkoski, Jr. 1984. Periodicity of extinctions in the geologic past. *Proc. U.S. Nat. Acad. Sci.* 81:801–805.

Rensch, B. 1959. *Evolution Above the Species Level.* New York: Columbia Univ. Press.

———. 1980. Neo-Darwinism in Germany. In *The Evolutionary Synthesis: Perspectives on the Unification of Biology,* ed. E. Mayr and W. D. Provine, 284–303. Cambridge, Mass.: Harvard Univ. Press.

Richardson, R. H. 1974. Effects of dispersal, habitat selection, and competition on a speciation pattern of *Drosophila* endemic to Hawaii. In *Genetic Mechanisms of Speciation in Insects,* ed. White M. J. D., 140–164. Sydney: Australia and New Zealand Book Co.

Ricklefs, R. E. 1979. *Ecology.* 2nd ed. New York: Chiron Press.

Riedl, R. 1978. *Order in Living Organisms.* New York: Wiley-Interscience.

Riska, B. 1981. Morphological variation in the horseshoe crab *Limulus polyphemus. Evolution* 35:647–658.

Rohde, K. 1979. A critical evaluation of intrinsic and extrinsic factors responsible for niche restriction in parasites. *Amer. Natur.* 114:648–671.

Rose, K. D., and T. M. Brown. 1984. Gradual phyletic evolution at the generic level in early Eocene omomyid primates. *Nature* 309:250–252.

Rose, M. R., and W. E. Doolittle. 1983. Molecular biological mechanisms of speciation. *Science* 220:157–162.

Rosen, D. E. 1978. Vicariant patterns and historical explanation in biogeography. *Syst. Zool.* 27:159–188.

———. 1979. fishes from the uplands and intermontane basins of Guatemala: Revisionary studies and comparative geography. *Bull. Amer. Mus. Nat. Hist.* 162:267–376.

Ross, H. H. 1972a. An uncertainty principle in ecological evolution. In *A Symposium on Ecosystematics,* ed. R. T. Allen and F. C. James 133–160. Occ. Paper no. 4. Fayetteville, Ark.: Univ. Arkansas Museum.

———. 1972b. The origin of species diversity in ecological communities. *Taxon* 21:253–259.

Ruse, M. 1982. *Darwinism Defended: A Guide to the Evolution Controversies.* Reading, Mass.: Addison-Wesley.

Saunders, P. T., and M. W. Ho. 1976. On the increase in complexity in evolution. *J. Theor. Biol.* 63:375–384.

———. 1981. On the increase in complexity in evolution. II. The relativity of complexity and the principle of minimum increase. *J. Theor. Biol.* 90:515–530.

Schrödinger, E. 1945. What is life? Cambridge: Cambridge Univ. Press.

Sekiguchi, K., and H. Sugita. 1981. Systematics and hybridization in the four living species of horseshoe crabs. *Evolution* 34:712–718.

Selander, R. K., S. H. Yang, R. C. Lewontin, and W. E. Johnson. 1970. Genetic variation in the horseshoe crab (*Limulus polyphemus*), a phylogenetic "relic." *Evolution* 24:402–414.

Senger, P. 1976. *Morphogenesis of Skin.* Cambridge: Cambridge Univ. Press.

Senger, P., and M. P. Patou. 1969. Experimental conditions in which feather morphogenesis predominates over scale morphogenesis. *Nature* 222:693–694.

Sewertzoff, A. N. 1931. *Morphologische Gesetzmassigkeiten der Evolution.* Jena: Fischer.

Shannon, C. E., and W. Weaver. 1949. *The Mathematical Theory of Communication.* Urbana: Univ. of Illinois Press.

Shields, W. M. 1982. *Philopatry, Inbreeding, and the Evolution of Sex.* Albany: State Univ. of New York Press.

Shimizu, H., and H. Haken. 1983. Co-operative dynamics in organelles. *J. Theor. Biol.* 104:261–273.

Simpson, G. G. 1944. *Tempo and Mode in Evolution.* New York: Columbia Univ. Press.

———. 1947. The problem of plan and purpose in nature. *Sci. Monthly* 64:481–495.

———. 1953. *The Major Features of Evolution.* New York: Columbia Univ. Press.

Slatkin, M. 1981. Estimating levels of gene flow in natural populations. *Genetics* 99:323–335.

Sneath, P. H. A., and R. R. Sokal. 1973. *Numerical Taxonomy.* San Francisco: W. H. Freeman.

Sober, E. 1975. *Simplicity.* Oxford: Clarendon Press.

Stebbins, G. L., Jr. 1971. *Chromosomal Evolution in Higher Plants.* Reading, Mass.: Addison-Wesley.

Stebbins, G. L., and F. J. Ayala. 1981. Is a new evolutionary synthesis necessary? *Science* 213:967–971.

Sturtevant, A. H. 1919. *Contributions to the Genetics of Drosophila melanogaster.* III. *Inherited Linkage Variations in the Second Chromosome,* 305–341. Publication 278. Washington, D.C.: Carnegie Institute.

Suzuki, D. T. 1973. Genetic analysis of crossing-over and its relation to chromosome structure and function in *Drosophila melanogaster.* In *Genetic Lectures,* ed. R. Bogard, 7–32, vol. 3. Corvallis: Oregon State Univ. Press.

Tax, S., ed. 1959. *Evolution After Darwin.* Vol. I, *The Evolution of Life.* Chicago: Univ. of Chicago Press.

Templeton, A. R. 1981. Mechanisms of speciation—a population genetic approach. *Ann. Rev. Ecol. Syst.* 12:33–48.

Thoday, J. M., and T. B. Boam. 1959. Effects of disruptive selection. II. Polymorphism and divergence without isolation. *Heredity* 13:205–218.

Thomerson, J. E. 1966. A comparative biosystematic study of *Fundulus notatus* and *Fundulus olivaceous* (Pisces: Cyprinodontidae). *Tulane Stud. Zool.* 13:29–47.

Thomerson, J. E., and D. P. Wooldridge. 1970. Food habits of allotopic and syntopic populations of the topminnows *Fundulus olivaceous* and *Fundulus notatus. Amer. Midl. Nat.* 84:573–576.

Treadwell, A. L. 1898. Equal and unequal cleavage in annelids. *Biol. Lect. MBL (Woods Hole)* 1898:93–111.

Tribus, M. 1961. Informatin theory as the basis for thermostatics and thermodynamics. *J. Appl. Mech. Trans. ASME* ((series E) 83:1–8.

———. 1983. Thirty years of information theory. In *The Study of Information,* ed. F. Machlup and U. Mansfield, 475–484. New York: Wiley-Interscience.

Turing, A. M. 1952. The chemical basis of morphogenesis. *Philos. Trans. Royal Soc. Lond. B* 237:37–72.

van der Steen, W. J., and B. Voorzanger. 1984. Methodological problems in evolutionary biology III. Selection and levels of organization. *Acta Biotheor.* 33:199–213.

Van Ness, H. C. 1983. *Understanding Thermodynamics.* New York: Dover.

von Baer, K. E. 1828. *Uber Entwicklungsgeschichte der Thiere.* Konigsberg: Borntrager. Part 1.

von Bertalanffy, L. 1933. *Modern Theories of Development: An Introduction to Theoretical Biology.* London: Oxford Univ. Press.

———. 1952. *Problems of Life: An Evaluation of Modern Biological Thought.* London: Watts.

Vrba, E. S. 1980. Phylogenetic analysis and classification of fossil and recent Alcelaphini (Mammalia: Bovidae). *Biol. J. Linn. Soc.* 11: 207–228.

———. 1984. What is species selection? *Syst. Zool.* 33:318–328.

Waddington, C. H. 1957. *The strategy of the genes.* London: Allen and Unwin.

———. 1966. *New Patterns in Genetics and Development.* New York: Columbia Univ. Press.

———. 1977. *Tools for Thought.* St. Albans: Paladin.

Wagner, W. H., Jr. 1961. Problems in the classification of ferns. In *Recent Advances in Botany,* 841–844. Montreal: Univ. of Toronto Press.

———. 1983. Reticulistics: The recognition of hybrids and their role in cladistics and classification. In *Advances in Cladistics,* ed. N. I. Platnick and C. J. Humphries, vol. 2, 63–79. New York: Columbia Univ. Press.

Wahlund, S. 1928. Zusammensetzung von Populationen und Korrelationserscheinungen vom Stanpunkt der Verebungslehre aus betrachtet. Hereditas 11:65–106. (Reference from Nei 1965.)

Watrous, L. E. 1982. [Review of] Phylogenetics: The Theory and Practice of Phylogenetic Systematics. Syst. Zool. 31:98–100.

Watrous, L. E., and Q. D. Wheeler. 1981. The out-group comparison method of character analysis. *Syst. Zool.* 30:1–11.

Webster, G. C., and B. C. Goodwin. 1982. The origin of species: A structuralist approach. *J. Social Biol. Struct.* 5:15–47.

Weichert, C. K. 1970. *Anatomy of the Chordates.* New York: McGraw-Hill.

Welch, B. L. 1963. From coral reef to tropical island via *Thalassia* and mangrove. *Va. J. Sci.* 14:213–214.

White, M. J. D. 1973. *Animal Cytology and Evolution.* Cambridge: Cambridge Univ. Press.

———. 1978. *Modes of Speciation.* San Francisco: W. H. Freeman.

Whitman, C. O. 1888. A contribution to the history of the germ-layers of *Clepsine. J. Morphol.* 1:105–182.

Wicken, J. S. 1979. The generation of complexity in evolution: A thermodynamic and information-theoretical discussion. *J. Theor. Biol.* 77:349–365.

———. 1980. A thermodynamic theory of evolution. *J. Theor. Biol.* 87:9–23.

———. 1981. Causal explanations in classical and statistical thermodynamics. *Philos. Sci.* 48:65–77.

Wigglesworth, V. B. 1954. *The Physiology of Insect Metamorphosis.* Cambridge: Cambridge Univ. Press.

Wiley, E. O. 1978. The evolutionary species concept reconsidered. *Syst. Zool.* 27:17–26.

———. 1980a. Is the evolutionary species fiction? A consideration of classes, individuals and historical entities. *Syst. Zool.* 29:76–79.

———. 1980b. Phylogenetic systematics and vicariance biogeography. *Syst. Bot.* 5:194–220.

————. 1981a. *Phylogenetics. The Theory and Practice of Phylogenetic Systematics*. New York: Wiley-Interscience.

————. 1981b. Convex groups and consistent classifications. *Syst. Bot.* 6:346–358.

————. 1981c. Remarks on Willis' species concept. *Syst. Zool.* 30:86–87.

Wiley, E. O., and D. R. Brooks. 1982. Victims of history—a nonequilibrium approach to evolution. *Syst. Zool.* 31:1–24.

————. 1983. Nonequilibrium thermodynamics and evolution: A response to Løvtrup. *Syst. Zool.* 32:209–219.

Williams, G. C. 1975. *Sex and Evolution*. Princeton: Princeton Univ. Press.

Williamson, P. G. 1981. Palaeontological documentation of speciation in Cenozoic molluscs from Turkana Basin. *Nature* 293:437–443.

Wilson, E. B. 1891. Some problems in annelid morphology. *Biol. Lect. MBL (Woods Hole)* 1890:53–78.

————. 1892. The cell-lineage of *Nereis*. A contribution to the cytogeny of the annelid body. *J. Morphol.* 6:361–480.

————. 1895. The embryological criterion of homology. *Biol. Lect. MBL (Woods Hole)* 1895:101–124.

Wilson, E. O., and W. H. Bossert. 1971. *A Primer of Population Biology*. Sunderland, Mass.: Sinauer Assoc.

Wright, S. 1931. Evolution in Mendelian populations. *Genetics* 16:97–159.

————. 1968. *Evolution and the Genetics of Populations*. Vol. 1, *Genetic and Biometric Foundations*. Chicago: Univ. of Chicago Press.

————. 1969. *Evolution and the Genetics of Populations*. Vol. 2, *The Theory of Gene Frequencies*. Chicago: Univ. of Chicago Press.

————. 1977. *Evolution and the Genetics of Populations*. Vol. 3, *Experimental Results and Evolutionary Deductions*. Chicago: Univ. of Chicago Press.

————. 1978. *Evolution and the Genetics of Populations*. Vol. 4, *Variability Within and Among Natural Populations*. Chicago: Univ. of Chicago Press.

Yamamoto, M. 1979. Interchromosomal effects of heterochromatic deletions on recombination in *Drosophila melanogaster*. *Genetics* 93:437–448.

Yockey, H. P., ed. 1958. *Proceedings of a Symposium on Information Theory in Biology*. New York: Pergammon Press.

Zar, J. H. 1984. *Biostatistical Analysis*. Englewood Cliffs, N.J.: Prentice-Hall.

Zotin, A. I. 1972. Thermodynamic Aspects of Developmental Biology. In *Monographs in Developmental Biology*, ed. A. Wolsky, vol. 5. Basle: S. Karger.

Zotin, A. I., and R. S. Zotina. 1978. Experimental basis for qualitative phenomenological theory of development. In *Thermodynamics of Biological Processes*, ed. I. Lamprecht and A. I. Zotin, 61–84. Berlin: de Gruyter.

Zuckerkandl, E. 1976. Programs of gene action and progressive evolution. In *Molecular Anthropology*, ed. M. Goodman, R. E. Tashian, and J. E. Tashian, 387–447. New York: Plenum Press.

· Index ·